North American Mixing Forum 2021

Held at the 2021 AIChE Annual Meeting

Boston, Massachusetts, USA and Online
7 - 11 November and 15 - 19 November 2021

ISBN: 978-1-7138-5702-0

Printed from e-media with permission by:

Curran Associates, Inc.
57 Morehouse Lane
Red Hook, NY 12571

AIChE
120 Wall Street, FL 23
New York, NY 10005-4020

Phone: (800) 242-4363
Fax: (203) 775-5177

www.aiche.org

Additional copies of this publication are available from:

Curran Associates, Inc.
57 Morehouse Lane
Red Hook, NY 12571 USA
Phone: 845-758-0400
Fax: 845-758-2633
Email: curran@proceedings.com
Web: www.proceedings.com

TABLE OF CONTENTS

621774 Turbulence Damping Above the Cloud Height in Stirred Tanks with 20-30 Weight Percent Glass Beads: Data for Multiphase CFD Simulations

Abstract:

Poor mixing in the clear liquid layer above the cloud height has been reported by several authors. This study uses LDV measurements to quantify turbulence above the cloud using a liquid level of 1.5 T to remove the barrier of a free surface at H = T. A D = T/3, down-pumping PBT was used at an off-bottom clearance of C = T/3. Three slurries were tested at impeller speeds 0.8, 1, and 1.2Njs. The change in turbulence was quantified using the normalized root mean square (RMS) of the fluctuating velocity summed and averaged over each radial traverse. A significant difference between the fluctuating values of the cloud height—minimum, average, and maximum—was observed. The turbulence decays until the maximum cloud height. Beyond that, it remains constant and near zero. The effects of both particle size and solids concentration prove to be important.

Note: This paper was recently published in the AIChE Journal but has not previously been presented.

624034 A Multiscale Lagrangian Analysis of Turbulence in Agitated Tanks

Abstract:

Mixing aims to enhance the homogeneity of a single-phase or multiphase system through input of mechanical energy to achieve a desired process result. This is an important industrial operation which is often conducted in a mechanically agitated vessel and is critical to the successful manufacturing of numerous products including fine chemicals, pharmaceuticals, personal/home care products, paper and pulp, polymers, food, and the formulation of products for these sectors. The design of mechanically agitated vessels, however, is still often as much an art as a science and for many applications it cannot be carried out from first principles. Therefore, theoretical methodologies for evaluation of mixing performance are crucial.

The process of mixing occurs at two different levels: micromixing and macromixing. Quantitative analysis at both levels is important for studying fluid mixing problems. Micromixing is achieved through molecular diffusion of individual species and eddy motion. Improper mixing on this level leads to segregation. Macromixing, on the other hand, relies on bulk or convective motion to distribute the materials to be mixed and reduce their local concentration gradients. These levels of mixing can be quantitatively assessed using decomposed Lagrangian trajectories of fluid elements or particles. On the basis of this principle, we exploit Lagrangian experimental data obtained by a unique technique of positron emission particle tracking (PEPT) to develop a new methodology for characterizing mixing levels in mechanically agitated vessels. In PEPT, radio-labelled particles are used as flow followers and tracked in 3D space and time through positron detectors. Compared with leading optical laser techniques (e.g. LDV, PIV), PEPT has the enormous and unique advantage that it can image opaque fluids, and fluids inside opaque apparatus with comparable accuracy. Thus, PEPT can provide the long-term 3D trajectories of all the components in a multiphase particle-liquid flow.

The new technique herein proposed is based on wavelet decomposition of Lagrangian trajectories provided by PEPT and is used to separate the mean (deterministic) and oscillatory (stochastic) parts of Lagrangian trajectories. The mean part includes low-frequency (non-diffusive; i.e., background motion) component while the oscillatory part represents high-frequency (diffusive; i.e., eddies) component of the trajectory. The implementation and potential of this new method are demonstrated by analysing long-term PEPT trajectories tracked in single-phase liquid inside a 30 cm diameter vessel agitated by a six pitched-blade turbine, operating in both up- and down-pumping configurations over a wide range of experimental conditions. Detailed information is obtained on turbulent kinetic energies as well as turbulent diffusion coefficients, allowing the assessment of mixing on both the micro and macro levels (Fig. 1). Such detailed information is invaluable for unravelling the complexities of single and multiphase flows inside stirred vessels including the ability to identify optimal cells for injection or withdrawal of suspension or other additives such as reactants.

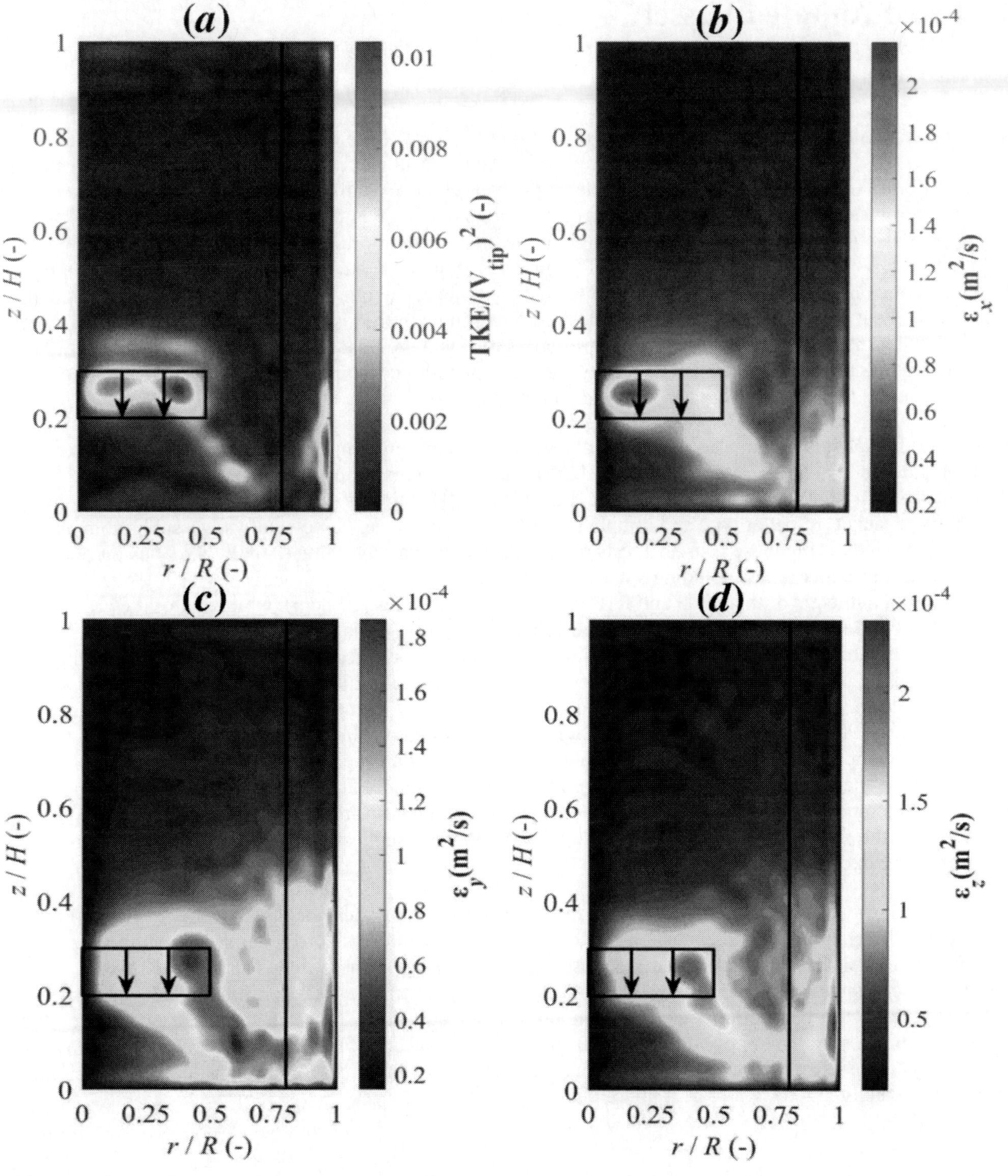

Figure 1. Azimuthally-averaged radial-axial maps of (a) normalized turbulent kinetic energy, (b-d) turbulent eddy diffusion coefficients in x, y and z-direction for a single-phase system at 260 rpm

624573 The Effect of Impeller Type on Heating Times in a Jacketed Agitated Vessel

Abstract:

A great many industries have a genuine need to control the temperature of a working media within an agitated vessel. This can be accomplished by installing an external jacket on the vessel sides, and / or the vessel bottom. Pumping a heat transfer fluid through a vessel jacket will add or remove heat from the working media as it mixes. The agitation system that circulates the working media promotes not only mixing, but also temperature uniformity throughout the media and more effective heat transfer between it and the vessel wall. Mixing literature (e.g., J.R. Nunhez, Chapter 14, Advances in Industrial Mixing, pp 491-532, John Wiley & Sons, New York, NY, 2016) contains numerous empirical correlations to calculate process-side heat transfer coefficients under desired operating conditions, as well as heating / cooling times of the working media given certain physical properties of it and the heat transfer fluid. These correlations indicate that process-side heat transfer coefficients are dependent upon impeller type, or perhaps more accurately, upon flow patterns created by the different impellers.

This experimental study compares the performance of three common impeller types while heating up a batch of water in a jacketed vessel. A total of 42 different heat transfer tests were completed at nine different sets of operating conditions in an 18.1inch (46cm) ID stainless steel vessel with dished bottom and a dimple jacket. Impeller type and dimple jacket surface area (i.e., vessel wall vs. vessel bottom vs. both) were varied during these tests. All tests mixed a constant water volume (21gal; 79.4L) using one of three different impeller types including a narrow hydrofoil (A310), a Rushton turbine (R100), or a 4-45-PBT (A200). Diameters of the A200 and R100 impellers are 7inches (17.8cm), while that of the cast A310 is 6.8inches (17.3cm), the nearest-to-the-standard diameter that is available in the cast A310. Each test monitored temperature in the agitated water using two vertically oriented thermocouples mounted 1inch (2.5cm) from the vessel wall, and either 2inches (5.1cm) below the liquid surface or 0.5inch (1.3cm) above the vessel bottom. A dedicated, PLC-controlled oil heater supplied heat transfer fluid to the vessel jacket at a temperature of 95-105deg F (35-41deg C). Each test simultaneously monitored increasing water temperature, decreasing oil temperature, oil flow rate, and total run time. A test lasted only as long as needed to raise the temperature of the agitated water to approximately 30deg F (17deg C) above the initial water source temperature (~55-60F; 13-15.8deg C). Each impeller was rotated at the requisite mixing speed to impart a fixed power level of 1.08 hp/kgal. Calculations suggest that heat up due to ongoing agitation never exceeded 2deg F (1deg C).

Test results show that differences between the two water-monitoring thermocouples were always within one degree of each other. Trends in heat up times with the different impellers generally match the trends predicted by correlations in the literature (e.g., see reference above). These trends also compare as expected when using either the vessel's dimple-jacket bottom or dimple-jacket wall but displayed some unexpected behavior when simultaneously using both. The results also showed a strong dependence on the behavior of the heating oil supply. For example, an accurate measure of oil flow rate was necessary to close the system energy balance, and fluctuations in oil flow rate and pressure drop produced unanticipated challenges in analyzing the water heat up trends.

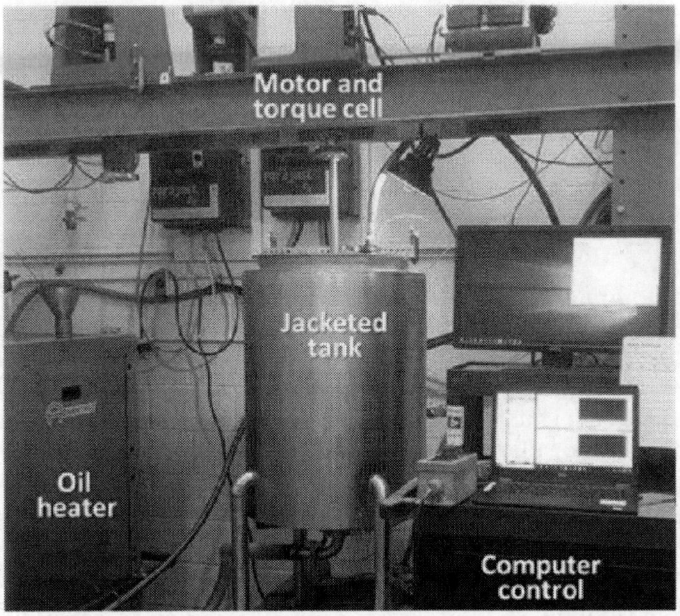

Laboratory system in which experiments were performed

Impellers used during laboratory experiments

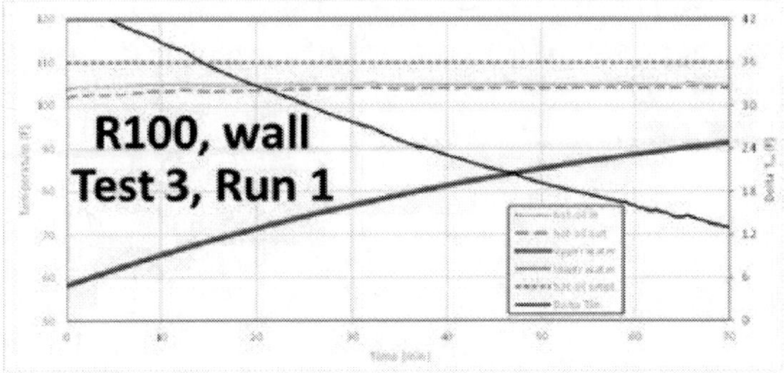

Typical heat up curve during laboratory experiments

627391 Comparing Predicted Just Suspended Speed Versus Experimental Just Suspended Speed for Axial Flow Impellers in a Stirred Vessel

Abstract:

Mixing applications that require suspension of solids in a liquid are common in just about every major industry. In many cases, the mixer just needs to keep the solids from settling, in order to allow the slurry to be transported from tank to tank. This is known as the just suspended condition, or the Zwietering criterion, which states that all particles have some vertical motion and do not remain on the bottom of the vessel for more than 1 to 2 seconds. Zwietering studied suspension of particles in stirred tanks to develop a correlation to predict the just suspended speed for a given impeller style (equation 1). This correlation has been an industry standard for decades. One limiting factor for the correlation is the impeller "S" constant, which is specific to the impeller style, impeller diameter to tank diameter ratio (D/T) and the impeller's location relative to the tank bottom. Thus, a specific "S" value is required for each geometry. Grenville, Mak and Brown developed a new correlation that brings in the impeller geometry (off bottom distance ratio) and categorizes impellers in two categories; pitched blade turbines and hydrofoils, since these are the most commonly used impellers for suspension applications. This correlation (equation 2) has similar elements to the Zwietering correlation [Advances in Industrial Mixing Chapter 10].

For this experimental study, a Lightnin A200 (pitched blade turbine) and a Lightnin A310 (hydrofoil) were tested over three volume scales to experimental determine the just suspended speed condition. Vertical tanks with four anti swirl wall baffles and a torispherical dished bottom were used for the study. The three scales were a 445 mm diameter vessel (90 liters), a 597 mm diameter vessel (220 liters) and a 800 mm diameter vessel (530 liters). A batch aspect ratio of 1.3 was used in each case with a single impeller near the bottom. The impellers were selected to match as closely as possible an existing geometry where an "S" value for the Zwietering correlation existed. Two types of glass beads were tested in each case; one with a particle size of 120 microns and one with a particle size of 300 microns all with a solids concentration of 1% wt. The mixer shafts were connected to strain gauge torque cells to measure shaft torque and optical tachometers to measure shaft speed. Although not a focus of this paper, cloud height was also measured for each condition tested. The Zwietering criterion was used to determine the just suspended speed condition by first increasing the speed of the mixer to achieve this condition and then reducing speed subsequentially to ensure the condition was not overshot.

At the time of writing this abstract the experimental procedure is partially finished. Initial results indicate that the experimentally determined just suspended speed is within +-10% of that predicted by each correlation. The work for the largest scale is not yet considered in this statement as that work is ongoing. A more detailed statistical analysis will be completed and presented.

$$(EQ-1):\ N_{js} = S\frac{\nu^{0.10} d_p^{0.20} X^{0.13} \left(\frac{g\Delta\rho}{\rho_L}\right)^{0.45}}{D^{0.85}}$$

$$(EQ-2):\ N_{js} = \frac{z}{N_p^{1/3} D^{2/3}} \left[\frac{g(\Delta\rho)}{\rho_L}\right]^{0.5} X_v^{0.154} d_p^{1/6} \left(\frac{C}{D}\right)^{0.1}$$

Variable	Description	Units
N_{js}	Impeller Speed for Just-Suspended State of Particles	rev/sec
S	Dimensionless Impeller Constant	No units
v	Kinematic Viscosity	m^2/s
g	Gravitational Constant	9.8 m/s^2
d_p	Particle Size	m
X	Mass Ratio of Suspended Solids to Liquid x 100 (%)	(kg solids / kg liquid) * 100
Δρ	Density Difference – Solids Density minus Liquid Density (ρ_s - ρ_L)	kg/m^3
ρ_L	Density of Liquid Phase	kg/m^3
D	Impeller Diameter	m
z	Dimensionless Constant in GMB Equation	No units
N_p	Impeller Power Number	No units
X_v	Volume of Solids per Unit Volume of Slurry x 100 (%)	Percentage

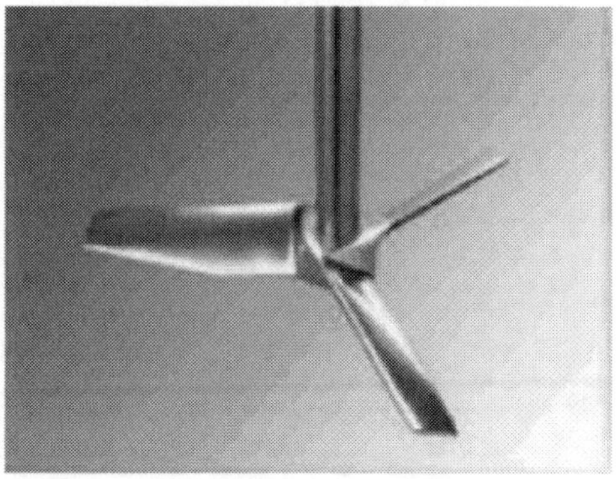

Figure 1: A310 Impeller

Figure 2: A200 Impeller

628291 Application of Electrical Resistance Tomography (ERT) to Solid-Liquid Mixtures in Stirred Vessels for the Determination of the Just-Suspended Speed, N_{js}

Abstract:

The minimum agitation speed to just-suspend finely dispersed solids in liquids in stirred vessels, often referred to as N_{js}, is a critical parameter to operate industrial mixing processes. A novel observer-independent method was developed using Electrical Resistance Tomography (ERT) in order to experimentally obtain N_{js} in vessels that cannot be visually inspected internally. Accordingly, the mean bulk resistivity was measured across electrodes mounted on a single ERT linear sensing probe placed vertically inside a stirred tank containing water and glass beads as the dispersed phase. As the agitation speed increased, more solids became suspended and the resistivity measured by the probe changed. Plots of resistivity variation vs. agitation speed resulted in an *S*-shaped curve, which was then mathematically analyzed by using several methods to extract the value of N_{js}. The N_{js} values obtained with this novel approach compared very favorably with those obtained using other methods requiring transparent tanks, namely the visual observation method originally developed by Zwietering (Zwietering TN. 1958. *Chem. Eng. Sci.* 8:244-253), and the imaging method recently developed by our research group (Teoman B, Shastry S, Abdelhamid S, Armenante PM. 2021. *Chem. Eng. Sci.* 231, 116263). It is expected that the novel ERT method proposed here could find wide applications in many industrial solid-liquid mixing processes.

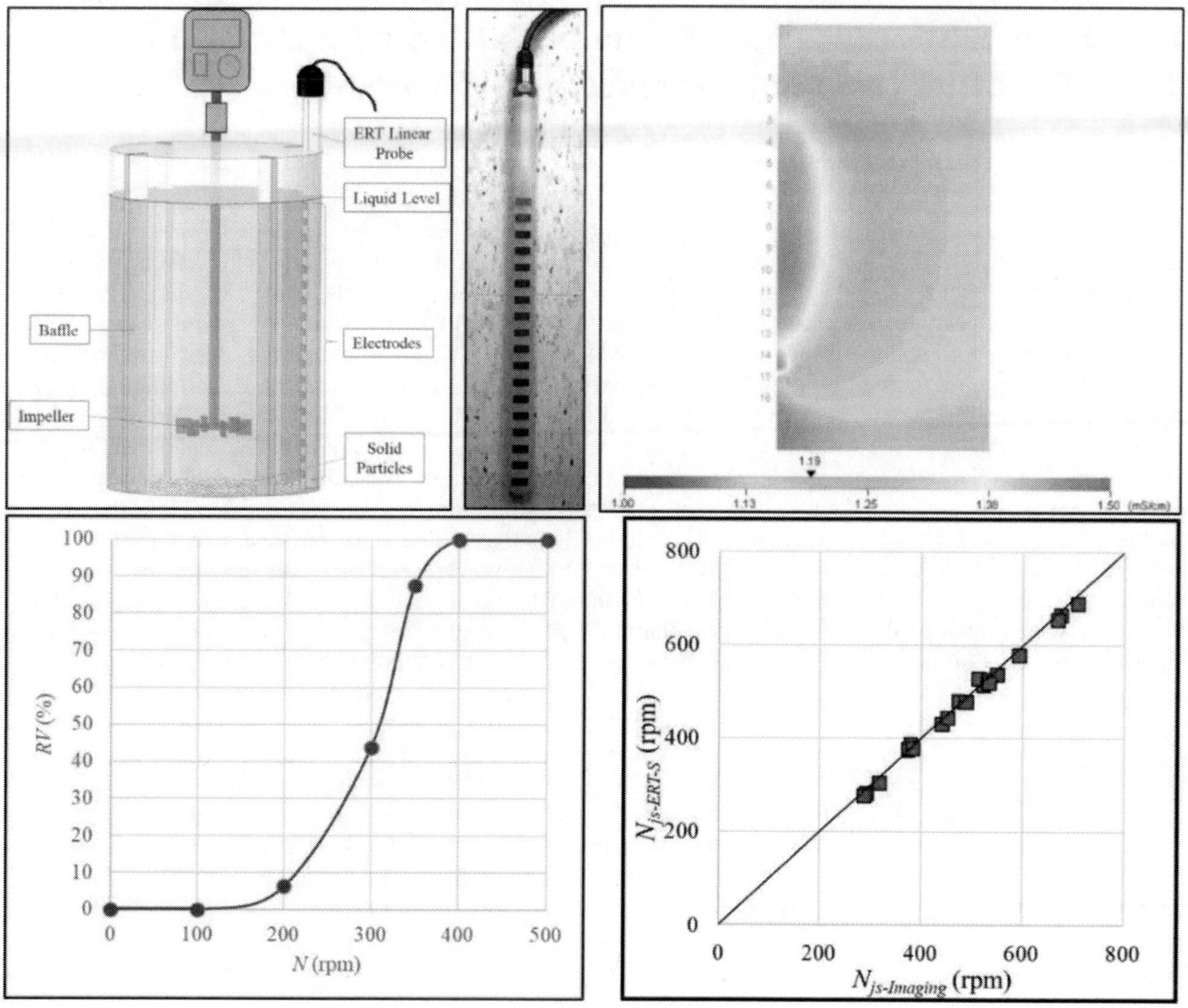
ERT Linear Probe
Liquid Level
Baffle
Electrodes
Impeller
Solid Particles
1.19
1.00
1.13
1.25
1.38
1.50 (mS/cm)
RV (%)
N (rpm)
0
10
20
30
40
50
60
70
80
90
100
0
100
200
300
400
500
N_js-ERT-S (rpm)
N_js-Imaging (rpm)
0
200
400
600
800

628696 Determination of the Power Number of a Retreat-Blade Impeller in a Stirred Vessel Partially Baffled with Different Types of Single Baffles Used in Pharmaceutical Processing

Abstract:

Mixing is an essential unit operation in the pharmaceutical and biopharmaceutical industry to conduct processes such as crystallization, fermentation, cell culture, chemical reactions, and others. In many such applications, the vessel/reactors typically utilized in processing plants are a glass-lined vessel provided with a single impeller, typically a Retreat-Blade Impeller (RBI), and a single baffle placed midway between the vessel wall and the impeller shaft. Since cleanability and chemical compatibility are critical factors, all internals, including the impeller and the baffle, must be glass-coated. In addition, the impeller is typically placed near the vessel bottom to avoid colliding with the baffle located in a more central position in the vessel rather than at the wall. While the presence and position of the single baffle make the system more easily cleanable and accessible, it also reduces the baffling action typically associated with fully baffled systems provided with four wall baffles. Therefore, single-baffle systems are often classified as "partially baffled." The most commonly used single baffle is the "beaver-tail" baffle consisting of a flattened elongated pipe closed at the end and glass-coated. However, over the years, equipment manufacturers have developed different types of single baffles with greater surface areas exposed to the rotating flow generated by the impeller greater than that of the beaver-tail baffle. Such types include the finger-type baffle, D-type baffle, H-type baffle, fin-type baffle and others types less commonly found. Despite their common use, very little information is available on the hydrodynamics of vessels equipped with these baffles and especially the impeller Power Number associated with their use. Thus, the objective of this study was to experimentally determine the power dissipation and Power Number in a scaled-down RBI system, using the various types of single baffles shown in the figure below, i.e., beavertail, D-type, fin-type, finger-type, and H-type baffles in the turbulent regime at different impeller agitation speeds. Computational Fluid Dynamics (CFD) was additionally used to predict the flow field in all these systems as well as the power dissipation and hence the Power Number. The use of different baffles resulted in the establishment of different flow patterns in the vessel. Furthermore, the CFD predictions and the experimental results for the Power Numbers were found to be in very close agreement. It is expected that this knowledge can help industrial practitioners in the optimization of pharmaceutical processes.

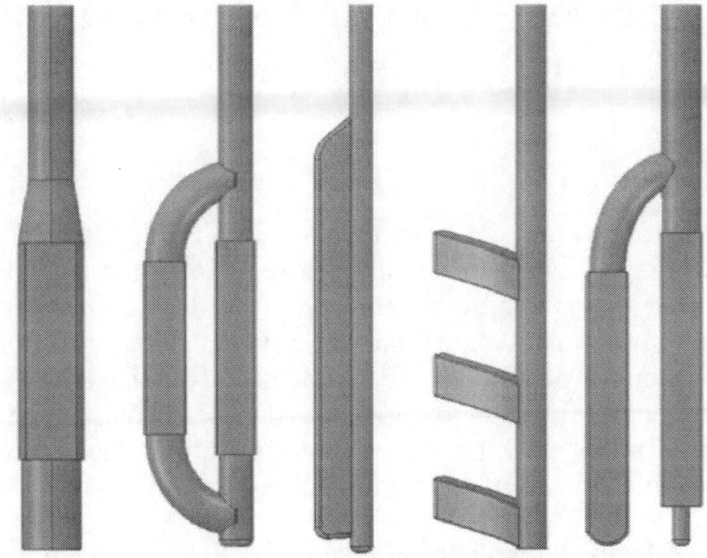

Beavertail Baffle | D-type Baffle | Fin-type Baffle | Finger-type Baffle | H-type Baffle

624706 An Exploratory Study of the Use of Combined Lattice Boltzmann – Discrete Element Method Modeling of Agitated Solid-Liquid Suspensions

Abstract:

The use and value of computational methods to understand and design agitated systems has been growing steadily. At this point, the use of simulations is routine, but methods continue to evolve and improve. The lattice Boltzmann approach to flow simulation and the discrete element method (DEM) of particle simulation are relatively new developments. Coupling these two tools provides an exciting opportunity for application to agitated solid-liquid suspensions. It is hoped that these approaches can provide more rigorous solutions without the need to adjust modeling parameters and to be a modeling expert. To be valuable for everyday application, this approach must provide accurate results with a reasonable amount of computational power and effort.

One of the most fundamental parameters to be provided by simulation of an agitated system is impeller power draw because of its impact on both process and mechanical design. Simulations currently provide accurate results of impeller power draw in liquid-only systems, but results in multiphase systems are lacking. Extensive experimental data that illustrate the effects of impeller type and solid characteristics on impeller power draw in these systems* provide the opportunity to critically evaluate the capabilities of simulation tools and determine if they are developed to the point that they can be successfully and confidently applied on a routine basis for agitator design.

Examples of experimental data that will be used to test the simulation results include the following:

- Addition of settling solids that increase suspension density have been found to increase the power number of all impellers (when power number is based on liquid density); however, the increase in power number with suspension density is less with flat impeller blades that generate large trailing vortices (e.g. – pitched-blade and Rushton turbines) than with profiled impeller blades that generate small trailing vortices (e.g. – hydrofoils).
- Addition of large neutrally-buoyant particles that do not alter suspension density would not be expected to affect impeller power draw. This lack of effect has been observed with a Rushton turbine, but the power draw of pitched blade turbines and narrow-blade hydrofoils increased dramatically with increasing solid content.
- Large settling solid particles increase the power number of a pitched-blade turbine more than small particles.

*K. J. Myers, E. E. Janz, T. Bao, and M. M. Heigl, "Impeller Power Draw During Turbulent Operation in Solid-Liquid Systems", Canadian Journal of Chemical Engineering, Volume 97, pages 2662 – 2670 (2019).

626549 Digitalise Process Assets for Improving Productivity: Enhance Process Insights, Metric and Parameters Via Simulations

Abstract:

Enhancing resource efficiency and decarbonisation are global priorities of manufacturing sector. There is a significant excitement about digitalisation, Industry 4.0 and their impact on manufacturing in process industries. It has a potential to become the most transformative era for chemical manufacturing (Industry 4.0: Top Challenges for Chemical Manufacturing, Elsevier White Paper, 2017). Digitalisation of process assets with the help of high-fidelity computational models is essential for realising this potential. However, development of computational models is only the first step towards it. It is essential to develop appropriate tools which will allow systematic digitalisation of process assets, derive new insights and synergistically evolve process understanding, key process metric and optimal process parameters for enhancing productivity. In this work, we present our approach towards developing such a tool for enabling enhanced productivity via digitisation of process assets.

Our approach uses science-based fundamental understanding as the focal point of process asset digitalization. Process understanding is most effectively represented using quantitative relationships between process parameters or PPs (such as impeller speed, liquid volume, solids loading etc.) and "Process Metrics" (PM). A PM is a (typically dimensionless) physical variable that describes some aspect of the physical process internal to a unit operation equipment [1]. For instance, in an agitated vessel bioreactor, the shear rates (and mass transfer coefficients) near the impeller is a PM - related to oxygen transfer. Similarly, the "dimensionless spray flux" in a high-shear wet granulator is a PM that relates to the effectiveness of the liquid binder spray with respect to the granulation process [2]. The aforementioned PP-PM relationships are the most appropriate means to digitally represent an asset. When a process asset has well defined PP-PM relationships, it is said to be "fully characterized" from a process performance perspective. For such digital characterization of performance assets, it is necessary to develop high-fidelity computational models. However, this is not sufficient. It is important to develop appropriate framework which goes beyond mere computational modeling and allows development of quantitative relationships between PP and PM and facilitates gaining new insights into process. In this talk, we present such a framework, database-driven web-based software framework - called "SimSight" which enables process insights via digital characterization of assets.

The web-based software framework, "SimSight" is generic in that it can be used in conjunction with any computational model. Here, we demonstrate this digitalization framework using computational fluid dynamics (CFD) models for agitated vessels - a widely employed class of process equipment. Other computational models such as discrete element method (DEM), Lattice Boltzmann method (LBM), chemical reaction engineering models etc. can be easily 'connected' to this framework.

SimSight consists of an asset database at its core - each asset being an object in the database. Attributes for each asset
include geometry information and simulation data/results. The simulation data/results are obtained via a "full computational characterization" of the asset. Such a characterization is enabled by running a set of simulations with process parameter settings dictated by a statistical DoE matrix. Using a set of calculation utilities provided along with this database of fully characterized process assets/equipment, SimSight empowers process engineers to create/share fundamental process knowledge in a more

streamlined way.

With its easy-to-use features, SimSight enables science-driven digitalization of process equipment and reduces time-to-scale for process development and transfer activities. We highlight key learnings of our development process. The application of SimSight is illustrated by a few case studies to bring out the inherent power and potential of such digitalization of process assets in improving productivity.

References

1. ICH Harmonized Tripartite Guideline, Pharmaceutical Development Q8 (R2)
2. Litster, J.D., K. P. Hapgood, J. N. Michaels, A. Sims, M. Roberts, S.K. Kamineni, T. Hsu, "Liquid Distribution in Wet Granulation: Dimensionless Spray Flux", Powder Technology 114 (1-3), pp 32-29

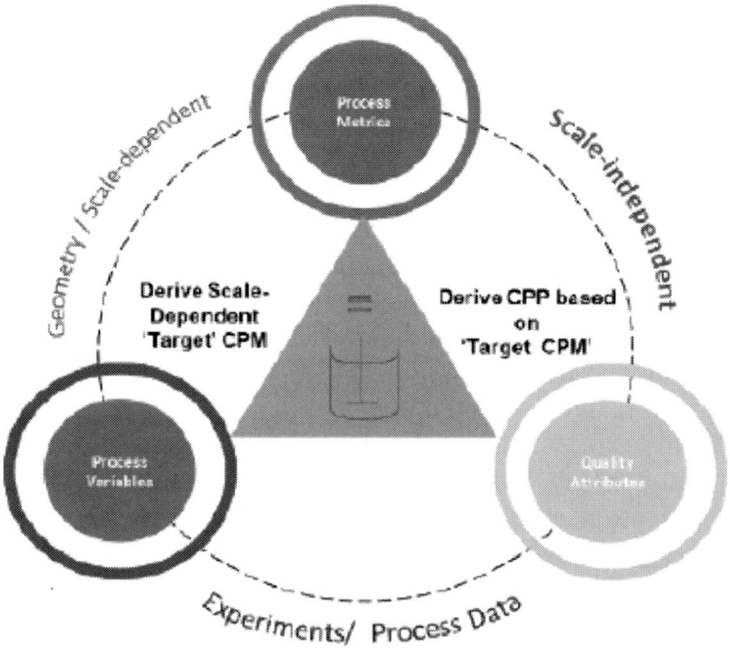

SimSight

628066 Using Hips As a New Mixing Model to Study Differential Diffusion of Scalar Mixing in Turbulent Flows

Abstract:

Mixing two or more streams is ubiquitous in chemical processes and industries involving turbulent liquid or gaseous flows. Modeling turbulent mixing flows is complicated due to a wide range of time and length scales, and non-linear processes, especially when reaction is involved. On the other hand, in turbulent reacting flows, sub-grid scales need to be resolved accurately because they involve reactive and diffusive transport processes. Transported PDF methods use mixing models to capture the interaction in the sub-grid scales. Several models have been used with varying success. In this study, we present a novel model for simulation of turbulent mixing called Hierarchical Parcel Swapping (HiPS). The HiPS model is a stochastic mixing model that resolves a full range of time and length scales with the reduction in the complexity of modeling turbulent reacting flows. This model can be used as a sub-grid mixing model in PDF transport methods, as well as a standalone model. HiPS can be applied to transported scalars with variable Schmidt numbers to capture the effect of differential diffusion which is important for modeling scalars with low diffusivity like soot. We present an overview of the HiPS model, its formulation for variable Schmidt number flows, and then present results for evaluating the turbulence properties including the scalar energy spectra, the scalar dissipation rate, and Richardson dispersion. These model developments are an important step in applying HiPS to more complex flow configurations.

631417 A Mechanistic Approach to Predicting Oxygen Mass Transfer in Bioreactors.

Abstract:

Stirred tank bioreactors are widely used in biopharmaceutical manufacturing unit operations. The quality of the drug substance produced during the operation is a function of the fluid mechanical environment realized within the vessel (e.g. dissolved gas concentrations, nutrient availability, and fluid shear rates etc.). In principle, this environment is a direct function of the vessel operating conditions. In practice, however, drug production is a complex convolution of the impeller speed, gas flow rate, fluid fill level, and reaction chemistry. Understanding the link between these operating conditions and the resulting environment inside the vessel is key to optimizing production. This understanding is particularly relevant to new bioreactor systems, where are not yet well characterized.

In this presentation, we propose a modeling capability in which hydrodynamics of a bioreactor along with its capability in transferring oxygen to the working fluid is assessed and validated against a set of published experimental data in literature. Specifically, we employ first principle mechanistic modeling to model oxygen mass transfer in a widely used lab-scale bioreactor (AMBR 15). For the first time, to the best of our knowledge, we show how this application allows the prediction of oxygen mass transfer to the fluid from both the open surface and sparger. This methodology is based on sound first principles and thus is not bioreactor specific, in other words, it can be applied across a wide range of scales and bioreactors.

633198 An Open-Source Adaptive and Distributed Nitsche Immersed Boundary Method for the Simulation of Laminar, Transitional and Turbulent Mixing

Abstract:

Computational Fluid Dynamics (CFD) has demonstrated that it can be a powerful tool to simulate the dynamics of a mixing system for both single and multiphase scenarios. In single phase mixing, it can be used to predict the residence time distribution, the mixing time, the occurrence of dead zones or the power consumption. This has led to a more widespread use of CFD in the mixing community.

The simulation of mixing system by CFD remains challenging because of the presence of a rotating impeller. The simplest approach is to formulate the CFD problem in a rotating frame of reference, but this breaks down for cases where the system does not present rotational symmetry: when baffles are present or when the impeller is off-centered. Two alternative solution exist to this problem: sliding-mesh strategies or the use of immersed boundaries.

Sliding-mesh strategies are available in multiple commercial or open source software (e.g. OpenFOAM). Although they are reliable, it is difficult and time-consuming to generate good quality (e.g. high orthogonality) meshes for these types of geometries. Furthermore, they are very cumbersome to use for configurations in which there are multiple impellers, each rotating with their own velocities. This hinders a more widespread usage of CFD, especially when it must be combined with other tools such as topology optimization or neural networks.

Immersed boundaries (IB) are an alternative approach in which the impeller is not meshed with a conformal mesh but is instead represented by discrete or continuous forcing terms [1]. Multiple IB approaches exist and have demonstrated they capabilities. However, most of them are either only available in closed-source CFD code or in commercial software. Additionally, they often suffer from a degradation of the accuracy of the underlying CFD scheme.

In this work, we present a novel parallel IB strategy that uses Nitsche's method to weakly impose the boundary conditions associated with the impeller. This Nitsche's method is implemented in a high-performance open-source FEM software: Lethe [2]. Lethe uses a Galerkin Least-Squares (GLS) approach to solve implicitly and at arbitrary order the incompressible Navier-Stokes equation. By leveraging the p4est and the deal.II library, it has the capacity to carry out parallel simulations with dynamic mesh adaptation and load balancing. Hence, using a Kelly error estimator, the mesh can be dynamically adapted to changing flow patterns in the mixing system, without degrading the computational performance of the underlying scheme.

In this talk, we first present Lethe and the GLS formulation used therein. We introduce our novel parallel distributed Nitsche's method and how it can be used to impose the motion of the impeller starting from a simple surface or volumetric mesh of the impeller itself. This approach is then first verified on test cases that related to mixing, but for which analytical/reference solution for the torque or the forces are available. Then, it is used to study the mixing in baffled and unbaffled systems over a large range of Reynolds number that cover the laminar to the early turbulent regime. This demonstrates that the model is capable of robustly simulate mixing system with minimal user input, since mesh generation and adaptation can be made fully automatic.

Finally, we conclude by discussing future application of the Lethe software and of this Nitsche's approach, notably in the field of multiphase mixing and granular mixing.

[1] Blais, B., Lassaigne, M., Goniva, C., Fradette, L., & Bertrand, F. (2016). A semi-implicit immersed boundary method and its application to viscous mixing. *Computers & Chemical Engineering*, *85*, 136-146.

[2] Blais, Bruno, et al. "Lethe: An open-source parallel high-order adaptative CFD solver for incompressible flows." *SoftwareX* 12 (2020): 100579.

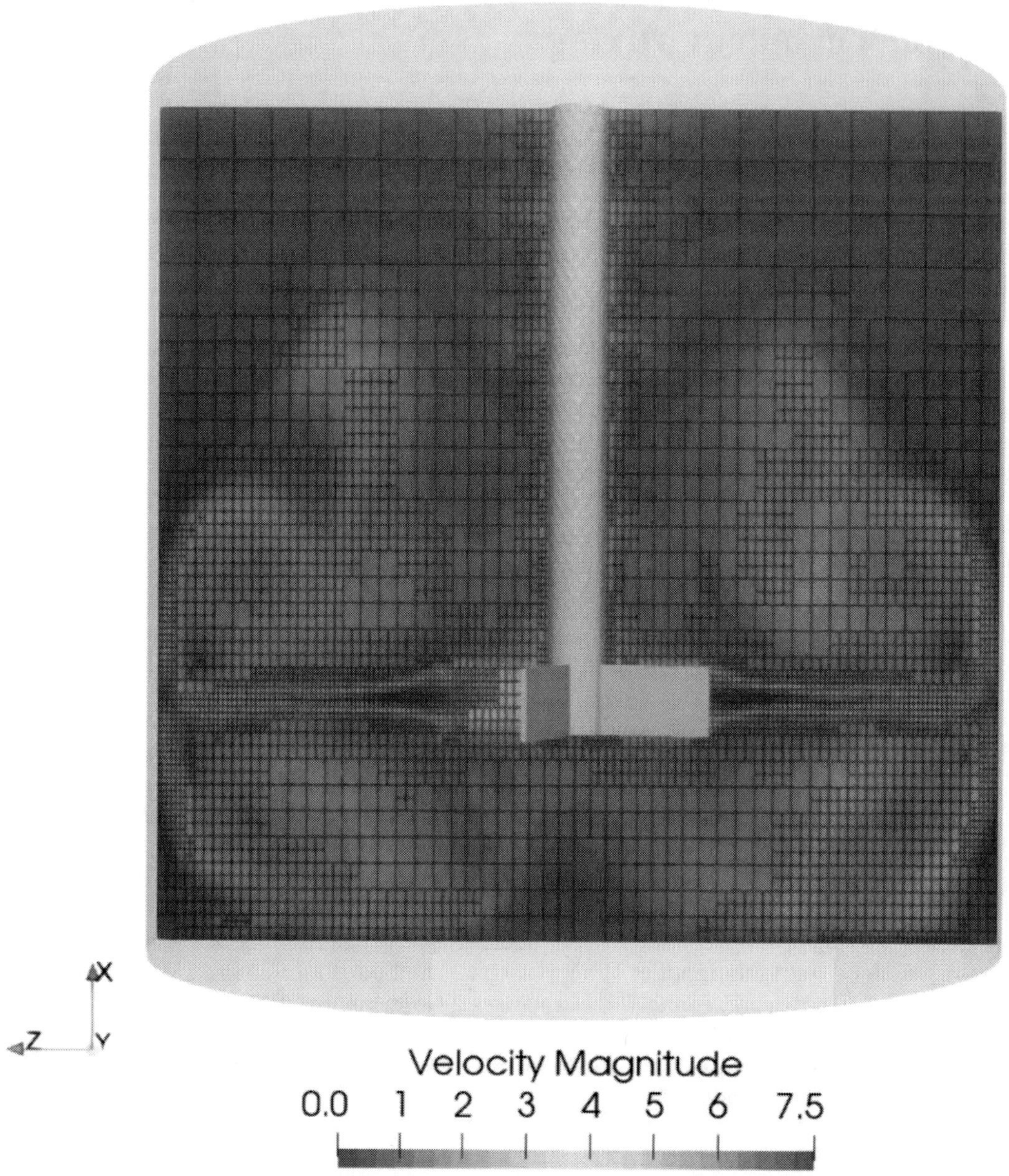

623463 Eulerian-Lagrangian Modelling of Particle-Liquid Flow and Mixing in a Stirred Vessel

Abstract:

Mechanically agitated vessels are widely used for various mixing operations within a wide range of industries including the chemical, pharmaceutical, food and petroleum industries. They are used for liquid blending, solid-liquid mixing, gas dispersion in liquids, heat/mass transfer enhancement and chemical reaction. Mixing is intrinsically a Lagrangian process and, whilst Eulerian data are essential, Lagrangian information is necessary for its complete description. Possible approaches of generating Lagrangian data can, in principle, employ numerical simulations or experimental techniques based on Lagrangian particle tracking to provide the trajectories of fluid elements or solid particles. We use computational fluid dynamics (CFD) simulations based on an Eulerian-Lagrangian approach in which particles are tracked through the continuous phase in transient mode, to predict the turbulent 3D flow of a particle-liquid suspension in a mechanically agitated vessel. Various computational codes are developed for performing Lagrangian statistical data analysis, Lagrangian-to-Eulerian data conversion, local pointwise multiphase mixing evaluation and detailed Eulerian plane by plane investigations inside the 3D flow. The CFD simulations are validated by comparing with experimental Lagrangian data obtained from a unique positron emission particle tracking (PEPT) technique which is able to accurately track different components of a multiphase flow in 3D space and time, and determine their individual long-term trajectories. The Eulerian velocity distributions inferred from both methods are in excellent agreement, confirming that the Eulerian-Lagrangian simulations are able to accurately predict the key features of the flow, as shown in Figures 1-6. These simulations are extended to study the flow of more complex multiphase particle-liquid suspensions relevant to industrial applications. The effects of various flow parameters are investigated including particle size, density and concentration, and carrier fluid properties. The aim of this work is to evaluate the capability of CFD to predict such complex flows and thus facilitate their modelling for research and industrial design purposes.

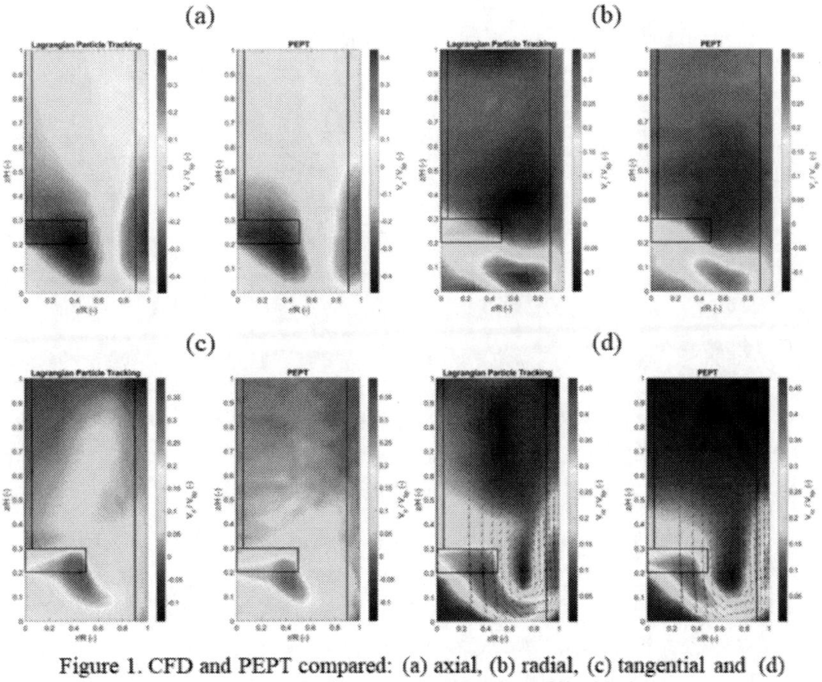

Figure 1. CFD and PEPT compared: (a) axial, (b) radial, (c) tangential and (d) mean axial-radial velocity distribution of liquid phase.

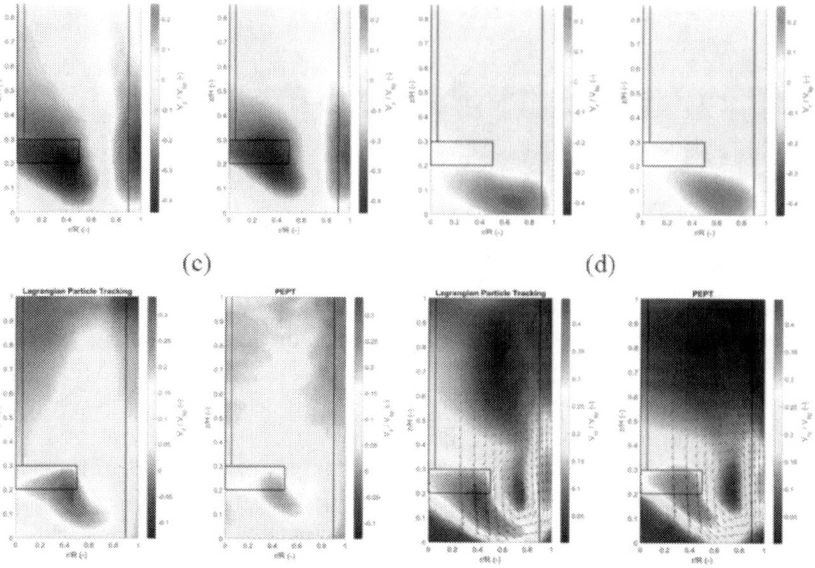

Figure 2. CFD and PEPT compared: (a) axial, (b) radial, (c) tangential and (d) mean axial-radial velocity distribution of solid phase.

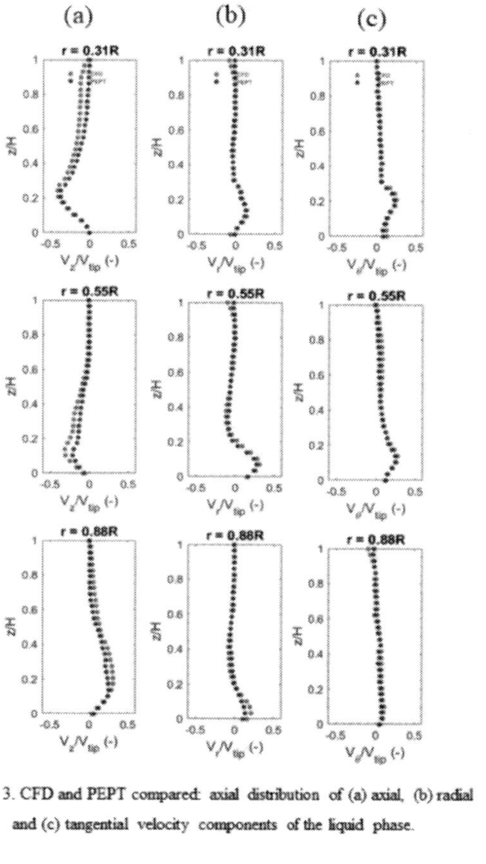

Figure 3. CFD and PEPT compared: axial distribution of (a) axial, (b) radial and (c) tangential velocity components of the liquid phase.

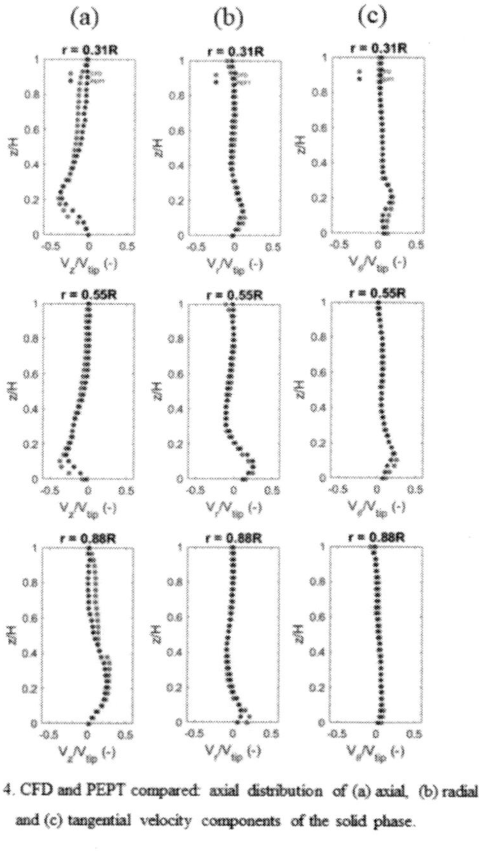

Figure 4. CFD and PEPT compared: axial distribution of (a) axial, (b) radial and (c) tangential velocity components of the solid phase.

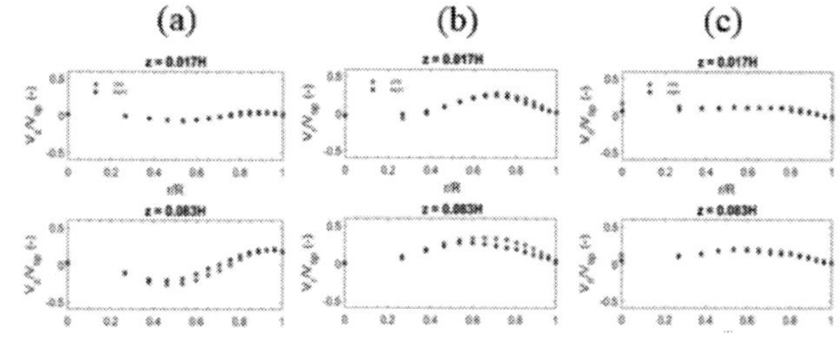

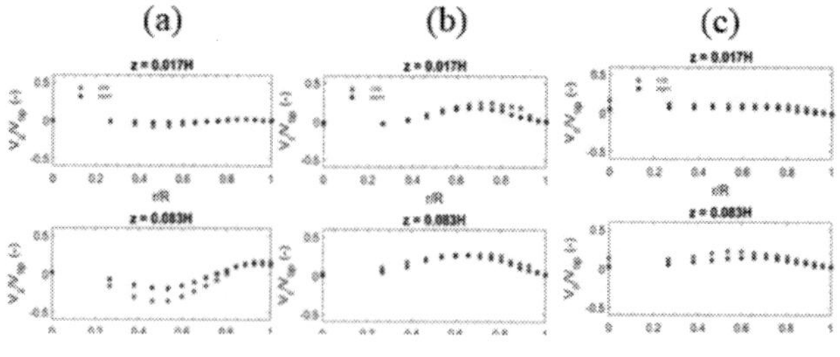

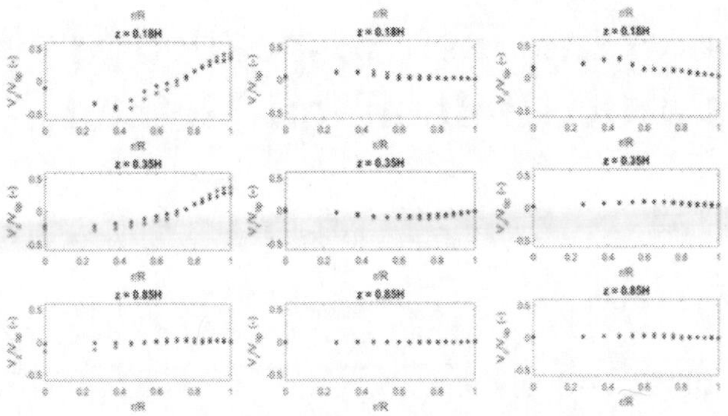

Figure 5. CFD and PEPT compared: radial distribution of (a) axial, (b) radial and (c) tangential velocity components of the liquid phase.

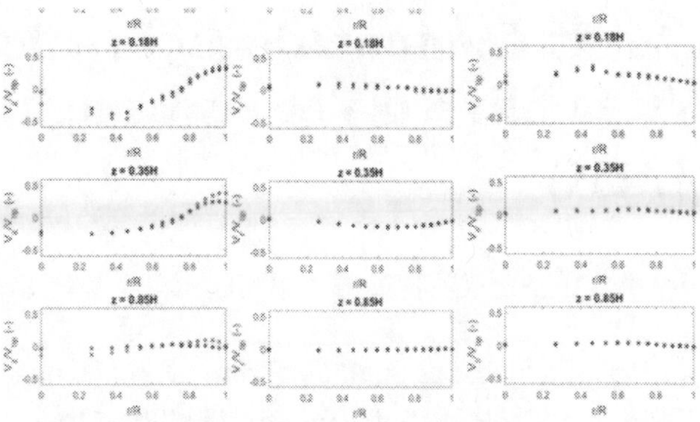

Figure 6. CFD and PEPT compared: radial distribution of (a) axial, (b) radial and (c) tangential velocity components of the solid phase.

626288 Computational Prediction of the Just-Suspended Speed, N_{js}, in Stirred Vessels Using the Lattice Boltzmann Method (LBM)

Abstract:

Industrial processes such as chemical reactions, dissolution of solids, and crystallization that involve finely dispersed solid particles suspended in a liquid in stirred vessels are widely used in the chemical and pharmaceutical process industries. Suspending all solids in the surrounding liquid is a crucial process objective in process optimization. The minimum agitation speed required to achieve off-bottom solid suspension, N_{js}, is a critical parameter used to ensure solid suspension in stirred vessels. Previous work has been extensively focused on the experimental determination of N_{js} using approaches that include solids sampling, visual observation, pressure gauge technique, steady cone radius measurements, and, more recently, electrical resistance tomography (ERT). Although each technique has different advantages and disadvantages such as objectivity, unobtrusiveness, visual observation, and applicability to industrial stirred vessels, it is still difficult and impractical for a wide range of industrial applications. Recently, Computational Fluid Dynamics (CFD) has been used to attempt to predict N_{js}, mainly using the Reynolds Average Navier Stokes (RANS) to simulate turbulence. However, solid suspension is three-dimensional and highly dependent on velocity fluctuations that are space and time dependent. It is thereby essential to introduce a more suitable CFD approach so as to predict N_{js} more reliably. This could additionally alleviate the need for extensive and expensive experimentation. Therefore, the objective of this study is to quantify N_{js} in a stirred tank provided with different axial and radial impellers using the Lattice Boltzmann Method (LBM). To do so, the mass fraction of suspended solid particles in a control volume near the bottom of the vessel was computationally predicted over a wide range of agitation speeds. Logistic regression analysis was combined with the computational approach to interpret the measurement data and hence predict N_{js} using a mathematical approach recently developed by our group to extract N_{js} from experimental ERT data. We validated the our CFD approach by comparing the predicted N_{js} results with available experimental data. The predicted N_{js} results were found in good agreement with those previously published, indicating that this CFD model combined with the logistic regression analysis proposed here are suitable for the determination of N_{js}. In turn, this work could help facilitate the solid suspension design and scale-up of industrial stirred vessels.

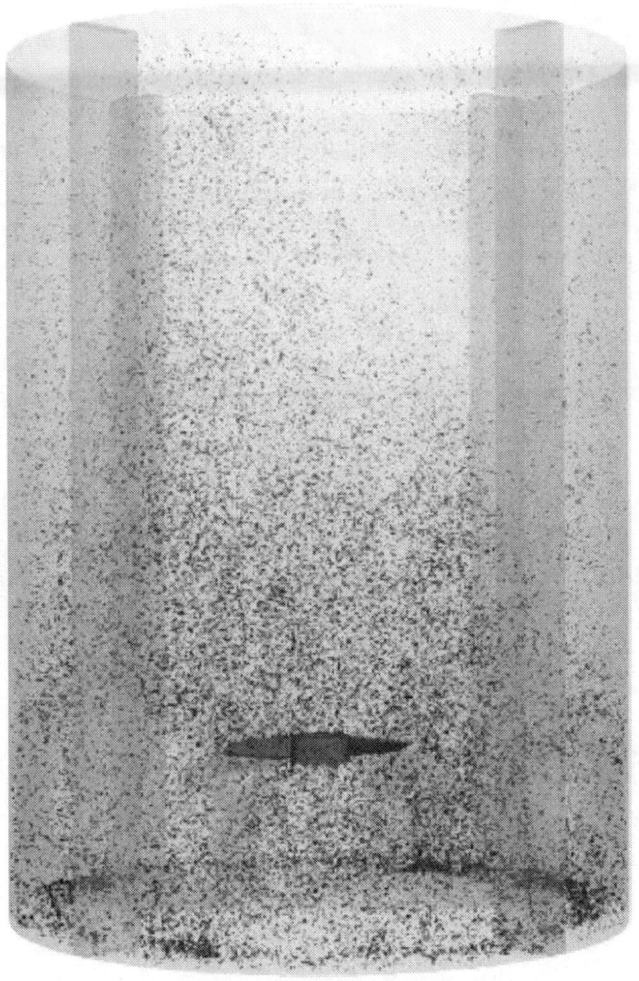

630277 CFD Analysis of Sparger Design and Bubble Formatoin, Breakup, and Coalesence.

Abstract:

In liquid phase reactions with gas sparging, both sparger design and agitation influence bubble area and mass transfer between the gas and liquid phases. In bubble columns, sparger design is the only influence; in stirred tanks, agitator design dominates. However, even in stirred tanks, non-uniform flow from the sparger can substantially impact the process. We show that modeling the sparger flow as a single phase system can predict non-uniform flow but ignores two phase interactions and fails to predict anything about bubble sizes. Using a fully resolved, two-phase Lattice Boltzmann method, where the interface between gas and liquid is explicitly modeled, we look at uniformity of flow in sparged systems including the effects of gas and liquid interactions. We look at breakup and coalescence of gas bubbles based on gas flow rate and agitation in the fully . We compare this to Eulerian-Lagrangian models where the gas phase is modeled as spherical bubbles with breakup and coalesence kernels based on dimensional analysis, which have previously been shown to predict mass transfer coefficients between gas and liquid within about 20% of experimental results. We also test the ability of these models to predict gas flooding in a Rushton turbine. We show that capturing coalesence behavior is essential in predicting the power drop in an agitated system upon flooding.

630930 Elucidating Design and Operational Impact on Dissolution Kinetics in Mixing Vessels through CFD Modelling

Abstract:

Introduction

In the manufacture of many pharmaceutical products, the essential unit operations are designed on a lab-scale and then are used to scale-up. The dissolution of solids in a liquid phase is one of these essential unit operations. Conventionally, these unit operations have been subjected to empirical techniques rather than numerical approaches to be optimized during the scale-up process.

This has changed in recent years though. Many studies have been focusing on analyzing the hydrodynamics of the agitated vessels with numerical methods. This includes varying the flow regimes and minimizing the vortex formation to better mix and enhance dissolution processes. To be able to model these phenomena through Computational Fluid Dynamics (CFD) there are several turbulence models and different model approaches to consider the impeller rotation and resolve the hydrodynamics present in the agitated vessels [1].

Many investigations have provided sufficient evidence that modeling solid-liquid mixing through CFD can have great advantages [2], [3], [4], [5], [6], [7]. Numerous configurations and arrangements can be modelled in less time and with a high accuracy. These advantages can lead to better scale-ups while reducing experimental costs [8]. However, dissolution kinetics are not explicitly included in these studies which are an important parameter to see the impact of underlying hydrodynamics.

The objective of this investigation is to combine the CFD hydrodynamics with the dissolution kinetics in an agitated vessel and provide insights about impact of design and operational configuration on the dissolution. To accomplish this, at first a CFD model is developed using Euler-Euler multiphase model together with dissolution kinetics and is followed up by simulating several scenarios to elucidate the impact of design and operational configurations on dissolution.

Methods

A CFD model is developed to investigate the dissolution kinetics of solute (NaCl) in water when mixed in a stirred vessel with a 50L volume. The model is transient, incompressible and multiphase. A two-phase Euler-Euler model was used to model the solute transport in the water. The impeller rotation was described by the Multiple Reference Frame (MRF) model [1]. The model was developed in Ansys Fluent 2019 R1©.

The dissolution rate was included explicitly in the model. This rate is function of the equilibrium solubility of the solute, the specific surface of the particles, the concentration of the dissolved material, the diffusion constant in the liquid phase, and the relative velocities of the particles and the fluid [8]. The equation 1 describes the dissolution rate [8]:

Whereas, is the dissolution rate of one particle, is the molar mass, is the diffusion coefficient, is the particle surface, is the diameter, is the gradient between the local liquid concentration and is the saturation concentration in the specified liquid [8].

The *Sh* is Sherwood number and is a function of Reynolds and Schmidt numbers (eq. 2) [8]:

Six different scenarios are modelled to assess the impact of operational and design configurations on the dissolution rate. The base case was designed to have the initial results and behavior of the dissolution kinetics and serve as a base line to compare with the rest of the scenarios. This base case was done in the 50L volume stirred tank and the solute was simply placed at the bottom of the tank. Scenario 1 has the same volume as the base case and the same mass, but the solute was introduced from the top by a single inlet. Scenario 2 is almost identical to scenario 1 but the solute is introduced through two inlets instead of one.

In scenario 3 a solute (Glucose) with lower diffusion coefficient is modelled with the same configuration as of Scenario 1. Scenario 4 is the scale-up scenario going from the initial 50L volume to 1000L volume. The final scenario, scenario 5, has the same configuration as scenario 1 with the only difference of change of the impeller location in the mixing vessel. The summary of the scenarios can be seen in the table 1.

Figure 1 shows the geometry used for scenario 1. A polyhedral mesh of size 800k was created to solve the continuity equations and the kinetics involved in the dissolution process.

Results

Figure 3 shows the volume fraction of solutes at 2 s in different scenarios. Different distributions in these scenarios shows the impact of different flow conditions arising due to different configurations. These different configurations lead to different dissolution times and hence shows the importance of design and operational conditions on the eventual dissolution.

Figure 4 shows mass of the solute in the solid state and Figure 5 shows the conc. of solute in the liquid. It is visible, that the base case has a longer dissolution time compared to scenario 1 and 2. This shows solute introduction method has a significant impact on the dissolution rate. However, the scenario 1 and 2 have almost the same dissolution rate of the solute in the mixing vessel. This demonstrates that in such configurations, one or two inlets have a small impact on the dissolution time. This could be attributed to the fact that inlets are close to central vortex that is formed due to stirring. Scenario 3 has the longest time due to the change in solute (from NaCl to glucose) and this change will affect the dissolution process because of the difference in the diffusion coefficient. It takes at least 120 s more than in scenario 1 for the solute to be completely dissolved.

Scenario 5 shows how the dissolution of the mass of the solute is affected when mixer location is changed. In Scenario 5, it takes a longer time to dissolved with the same amount of energy and for the same size of vessel (approximately 40s more). This reveals that the mixer location is crucial for the hydrodynamics and dissolution kinetics in the mixing vessels.

For scenario 4 the mass of solute is linearly scaled-up to be able to compare it with scenario 1 and it is seen that it has different dissolution performance. In this case scenario 4 takes a longer time for the solute to be dissolved in comparison with scenario 1. This can be explained due to the non-linearity of the scale-up process. This conclusion is also supported by plotting the average concentration of the solute in the solvent as mentioned before. This provides insight of how the variation between scenarios is and how important is to study and model the scale-up of a dissolution process. The study of the scale-up system will bring insights of how to introduce the solute and how it will behave in the stirred reactor to obtain similar results as in lab-scale testing.

Conclusion

This study presented a developed CFD model that can describe the kinetics of a dissolution process. It shows how changes in solute, design or operation have a direct impact on the dissolution kinetics in a mixing vessel. It also shows the benefits of virtual piloting to scale-up because it will give confidence to the decision makers to take the correct steps to do it. With the calculated variables process engineers will be capable of comparing a multitude of stirring equipment and its parameters. This could lead to optimize systems without the high costs of physical piloting.

References

[1] Ansys, "Fluent Theory Guide," ANSYS, Inc, Release R2021.

[2] S. Hosseini, D. Patel, F. Ein-Mozaffari and M. Mehrvar, "Study of solid-liquid mixing in agitated tanks through computational fluid dynamics modeling," Industrial & Engineering Chemistry Research, vol. 49, pp. 4426-4435, 2010.

[3] P. M. Armenante and E. U. Nagamine, "Effect of low off-bottom impeller clearance on the minimum agitation speed for complete suspension of solids in stirred tanks," Chemical Engineering Science, vol. 53, pp. 1751-1775, 1998.

[4] J. J. Derksen, "Numerical simulation of solids suspension in a stirred tank," AIChE Journal, vol. 49, pp. 2700-2714, 2003.

[5] H. Hartmann, J. J. Derksen, C. Montavon, J. Pearson, I. S. Hamill and H. E. van de Akker, "Assesment of large eddy and RANS stirred tank simulation by means of LDA," Chemical Engineering Science, vol. 59, pp. 2419-2432, 2004.

[6] K. J. Myers and A. Bakker, "Solids suspension with up-pumping pitched-blade and high-efficiency impellers," Canadian Journal of Chemical Engineering, vol. 76, pp. 433-440, 1998.

[7] S. M. Shekhar and S. Jayanti, "CFD study of power and mixing time for paddle mixing in unbaffled vessels," Chemical Engineering Research and Design, vol. 80, pp. 482-498, 2002.

[8] T. Hörmann, D. Suzzi and J. G. Khinast, "Mixing and Dissolution Processes of Pharmaceutical Bulk Materials in Stirred Tanks: Experimental and Numerical Investigations," Industrial & Engineering Chemical Research, vol. 50, pp. 12011-12025, 2011.

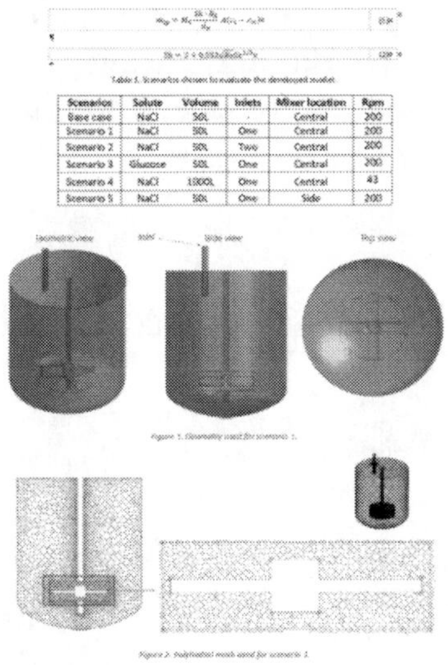

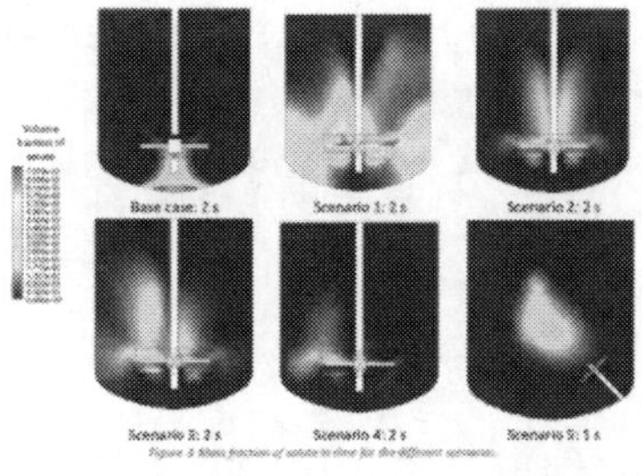

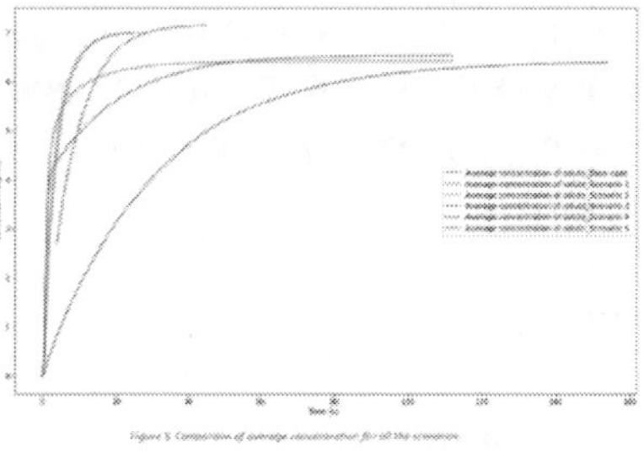

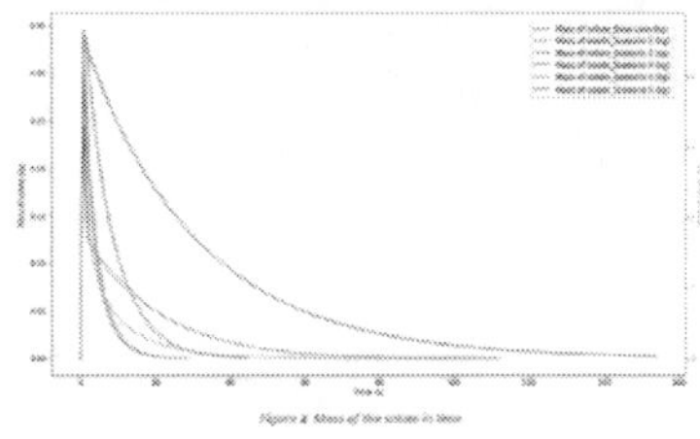

624189 A Mechanistic Approach for Predicting Mass Transfer in Bioreactors

Abstract:

The biomanufacturing processes that produce biologic drugs has become an extremely important area of
study within the pharmaceutical industry. Within such processes, the drug substance is typically produced
by living organisms within stirred tank bioreactors that require a continuous supply of sparged oxygen. The overall oxygen transfer rate to the fluid is a nonlinear convolution of the gas bubble size distribution,
fluid properties, local fluid energy dissipation rates, and local dissolved oxygen concentrations. The complexity of this process presents challenges to process scale-up and intensification. In this work,
we propose, implement, and validate a mechanistic transport model for predicting oxygen transfer rates
within stirred tank bioreactors. To begin, we describe the relevant conservation laws and key principles
from turbulence theory that govern mass transfer. Next, we present a physics-based modeling approach
for solving these equations in tandem and in real-time. We then systematically validate the model against experimental data at operating scales ranging from 5 L to 2000 L. By running the algorithm on graphics processing units (GPUs), the approach is shown to solve at timescales practical for industrial application.

Reference:

Thomas, John A., et al. "A Mechanistic Approach for Predicting Mass Transfer in Bioreactors." *Chemical Engineering Science* (2021): 116538.

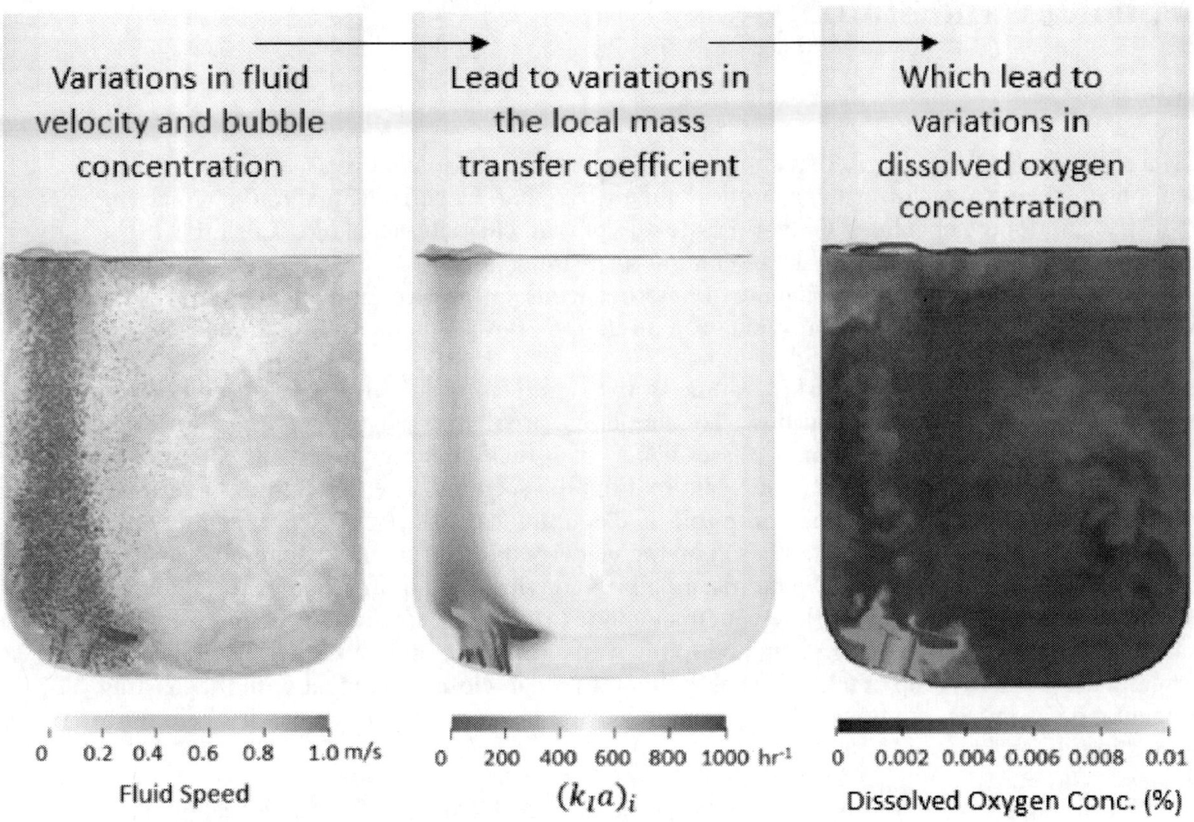

Variations in fluid velocity and bubble concentration
Lead to variations in the local mass transfer coefficient
Which lead to variations in dissolved oxygen concentration
0 0.2 0.4 0.6 0.8 1.0 m/s
Fluid Speed
0 200 400 600 800 1000 hr⁻¹
$(k_l a)_i$
0 0.002 0.004 0.006 0.008 0.01
Dissolved Oxygen Conc. (%)

631167 Effect of Surface Wettability Variation on Emulsion Morphology and Stability

Abstract:

Emulsions, liquid droplets dispersed in another immiscible liquid, are widely used in a myriad of applications in chemical engineering, analytical chemistry, pharmaceuticals, material engineering, oil and gas industry, and food technology. One of the most important characteristics of emulsion is its stability against separation. In this study, we carry out numerical simulations to investigate (oil-in-water) emulsion stability and morphology in a microchannel of heterogeneous surface wettability. The objective here is to elucidate the effect of surface wettability on drop motion via flow dynamics simulations.

We apply a conservative diffuse-interface lattice Boltzmann method (LBM) to perform three-dimensional numerical simulations of liquid-liquid flow. The numerical problem consists of a single (oil) drop immersed into an immiscible continuous phase (water) in a square microchannel with alternating surface wettability: hydrophilic, hydrophobic, and again hydrophilic sections. The two liquids have equal densities, and we vary the viscosity ratio between drop (oil) and continuous phase (water). Depending on the drop length and capillary number, calculated based on the velocity of the drop and dynamic viscosity of the continuous phase, four different flow patterns are observed when the oil drop passes through the hydrophobic section. These distinct flow patterns include (a) passing without any changes in drop shape or dynamics, (b) adhesion of the dispersed liquid to the walls, (c) phase inversion (i.e., water becomes the dispersed phase), and (d) drop breakage. These outcomes are in close agreement with the existing experimental data. The morphology of the oil emulsion and the corresponding velocity field for each case are shown in Fig 1.

Our simulation results reveal multiple recirculations present inside the oil droplet and within the surrounding thin-wetting film. Our preliminary data show that the thin-film recirculation tends to tear the film apart, leading to a change of emulsion morphology. These results suggest that the thin-film flow field plays a vital role in keeping emulsion stability.

In this presentation, we will elucidate how the emulsion stability and morphology under heterogeneous surface wettability are influenced by the following crucial factors: (i) the fluid viscosity ratio, (ii) surface wettability values, and (iii) the interfacial tension between the dispersed and continuous phases.

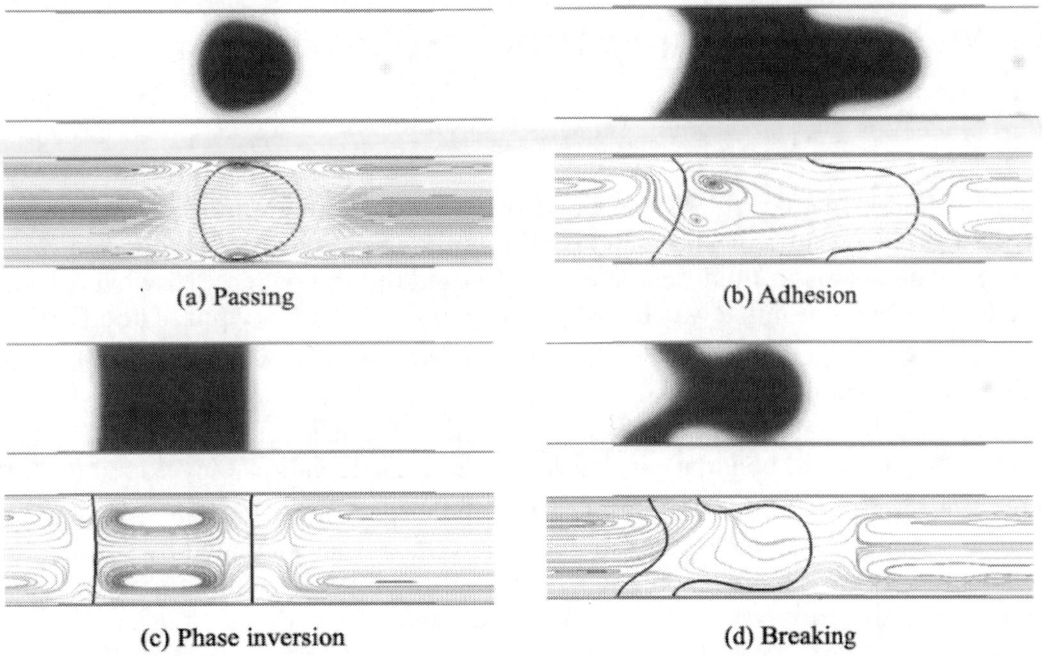

Fig. 1. Two-dimensional snapshots of phase-field (upper) and velocity streamlines (bottom) on the square microchannel's mid-plane. Depending on Capillary number and drop length, different emulsions dynamic is observed when the oil drop is passing the hydrophobic section (marked as red), encompassing (a) passing, (b) adhesion, (c) phase inversion, and (d) breaking, with the black curve representing the drop shape.

624194 Modeling Gas-Liquid Mass Transfer Using a Discrete Bubble Method: An Application to Bioreactors

Abstract:

Computational Fluid Dynamics (CFD) has been used in bioreactor engineering in general but most of the publications focus on studying single phase fluid mixing pattern and impeller power draw. These calculations provide details in fluid field and liquid-liquid mixing pattern. However, it has been challenging to effectively simulate gas bubble evolution as a secondary phase and O_2/CO_2 mass transfer phenomena that occurs on the gas-liquid interface around the bubbles. In this study, the physical process of mass transfer between the gas bubbles and bulk liquid is modeled using a Lattice-Boltzmann's approach i that significantly simplifies the multiphase model setup. The rate of mass transfer across the gas-liquid boundary is determined using correlations developed for turbulent flow that were previously reported in publications. The bubble specific surface area, [m^2/m^3], which indicates the area per unit volume of the gas-liquid interface was calculated by solving the two-phase fluid model using a discrete bubble approach. The resulting overall volumetric mass transfer coefficient was then determined.

The mass transfer calculation for both O_2 and CO_2 was performed for lab scale Applikon 3L bioreactors with two distinct impeller and sparger setups. The model predicted mass transfer rates based on simulated probe reading were compared to experimentally determined data from previous studies. The model predicted mass transfer rates of CO_2 and O_2 agrees well with experimentally determined values in both reactor setups over a wide range of impeller agitation rate and airflow rate. The model was also used to help understand the impact of Applikon 3 L reactor setup to fluid flow and mass transfer.

627369 CFD Modeling of Bioreactors: *Applying Euler-Lagrange LES Technique to Investigate Effect of Pressure & Sparge Air Flow on Dissolved CO_2 Profiles in a Fermentation process*

Abstract:

To understand the inhibitory effects of CO_2 toxicity in large scale fermentations it is critical to reliably predict dissolved CO_2 levels in industry scale fermenters. In this work, we attempt to apply a Lattice Boltzmann model (LBM) based Euler-Lagrange CFD approach to simulate the time-evolution of dissolved carbon dioxide and oxygen profiles in a fermentation process. The approach presented here eliminates the manual grid meshing process and runs orders-of-magnitude faster than conventional CFD modeling tools. Since this approach conserves energy, it directly predicts energy dissipation rates and turbulent energy spectrum at arbitrary points within the domain. We utilize this approach to simulate the mixing hydrodynamics, multi-phase flow behavior and multi-scalar transport of key (dissolved oxygen and carbon dioxide) species involved in the fermentation process. Full transient simulations have been attempted. Furthermore, through the utilization of CFD, parametric variations related to changes in headspace pressure and sparge air flow have also been captured on dissolved species profiles through well designed computational experiments. The key benefits of using the Euler-Lagrange approach over a conventional Euler-Euler framework includes reduced simulation run-times, and generality of the applied physics, without a need for parameter tuning.

630675 Computational Fluid Dynamics Simulation of Biomethanation in a Large-Scale Gas Bioreactor

Abstract:

Biomethanation (the process of methane formation by anaerobic microbes from hydrogen and carbon dioxide) at industrial scales is a promising solution to more sustainable and carbon-neutral hydrocarbon fuel production. However, gas fermentation at large scales have technical challenges that can unfavorably impact the productivity and operating costs. Low concentration of low-solubility dissolved gaseous components in the water, e.g., hydrogen in our case, can inhibit the fermentation if not thoroughly distributed throughout the reactor. On the other hand, if we oversupply substrate gas to the bioreactor, we will have unconsumed reactants that need to be captured and separated from the product gas (i.e., methane) at the gas exit line of the bioreactor. Reactor designs with optimal size and mixer positioning and actuation, gas sparging flow rates, mixer speed, etc., can alleviate these challenges and achieve high productivity and efficiency. Predictive computational fluid dynamics (CFD) simulations can significantly help with design and identification of optimal operating conditions for such bioreactors at large scales.

In this work, we have performed CFD simulations for a 30 L bioreactor. A previously developed multiphase Euler-Euler model [1] was modified to incorporate the additional gaseous species and multi-species reaction kinetics. This model is a customized solver derived from OpenFOAM [2], an open-source CFD toolbox. Our simulations include interphase mass transfer as well as species transfer between the phases. The microorganism activity is represented by simple phenomenological reaction kinetics for the following reaction:

$$CO_2 + 4\,H_2 \rightarrow CH_4 + 2\,H_2O$$

The dissolved hydrogen and carbon dioxide uptake rate and the methane production rate were incorporated in the kinetic reaction model. The mixer is a group of axially separated Rushton impellers whose parameters (number of blades, spacing between impellers, rotational speeds) can be varied at run time allowing for optimization of mixer design. The bioreactor is pressurized to 9 bar and preheated to 60 °C. We investigated the effect of gas flow rate and mixer speed on the performance of overall methane production and have also identified the regions in the bioreactor with low dissolved hydrogen and carbon dioxide concentrations. The simulation results will be used to help with the design of a future large-scale methane fermenting bioreactor.

References:

[1] Rahimi, M. J., Sitaraman, H., Humbird, D., and Stickel, J. J. (2018). Computational fluid dynamics study of full-scale aerobic bioreactors: Evaluation of gas–liquid mass transfer, oxygen uptake, and dynamic oxygen distribution. *Chemical Engineering Research and Design*, *139*, 283–295.

[2] Weller, H., Tabor, G., Jasak, H. and Fureby, C., A tensorial approach to computational continuum mechanics using object-oriented techniques. Computers in physics 12, no. 6 (1998): 620-631.

634072 Oxygen Transfer in Bioreactors with High Cell Densities

Abstract:

There is an urgent need to increase the production of antibody drug products manufactured from mammalian cells. This need has motivated a more rigorous analysis of existing bioreactor systems, and whether they can be extended to support order-of-magnitude step-increases in cell density and process yield. Systems with high cell densities present a unique challenge to process design and operation. Unlike traditional cell growth processes, which have modest gassing and oxygen transfer requirements, systems with high cell densities require oxygen transfer rates that can exceed 30 1/hr and gas flow rates in excess of 0.1 VVM. These high flow rates make *a priori* design and scale-up of such systems difficult, since most bioreactor design correlations are tailored for systems with more modest gassing rates.

In this work, we use lattice-Boltzmann simulation with Lagrangian bubble tracking to develop a mechanistic model for predicting the behavior of these high cell density bioreactors. All physics are coupled in space and time and mechanistically motivated by first principles. The bubbles evolve according to Newtons' second law using the local and instantaneous fluid velocity to compute a drag force. The fluid evolves according to the transient Navier-Stokes equations. The instantaneous fluid force from the bubbles on the fluid are incorporated as a body force. The species field is solved according to the advection-diffusion equation using the instantaneous fluid velocities. The bubbles act as species sources and sinks. In parallel with validation against experimental data, we use this model to examine the physics driving oxygen transfer within such high cell count systems. We then use the model to show how the operating conditions of the reactor can be tuned to achieve ideal oxygen transfer rates across the vessel.

630517 Effect of Tracer and Fluid Behavior on Mixing Performance of High Shear Mixers

Abstract:

High shear mixers (HSMs) broadly used in industrial to biological processes, such as cosmetics, paints, food, healthcare, coatings, etc., where intense rate of mixing, uniform emulsification and dispersion of fluids play a significant role (Vashisth et al. 2021). The flow fields and mixing in high shear mixers are intensified by applying high shear through narrow gaps between the rotor and stator geometries. In this study, the mixing performance for Newtonian and non-Newtonian fluids is analyzed for different HSMs. Three-dimensional numerical simulations are carried out for different stator heads HSMs such as disintegrated (18 stator holes), mesh (63 stator holes) and slotted (9 stator holes) at different rotor speeds (4000, 6000 and 8000 rpm). Standard k-ε turbulence (Vikash et al. 2017) and Power law (Zhang et al. 2017) models are used for the investigation of fluid dynamics of Newtonian and non-Newtonian fluids. Further, species transport model without reaction model is used for mixing characterization and simulations are run for flow time of 40 seconds for mixing characterization. The overall results suggested that mixing is a function of design of stator head, fluid behavior, rotor speed and tracer location (Figure 1(ii) and (iii)). In smaller stator holes HSM, cavern effect has been observed for viscous fluids and no improvement in the mixing is observed after certain flow time. Mixing time (t_{95}) is estimated for different HSMs for Newtonian fluid and a good agreement is found with the existing findings. For larger (slotted head) and smaller stator (mesh head) holes HSMs the mixing parameter constant (a) are found as 20.33 and 31.21, respectively, which is in line with the results available in the literature (James et al. 2017). However, the mixing constant is more than the stirred impellers (Grenville 1992). Further, it is observed that the tracer injection location has significant effect on the mixing performance. Tracer is injected at three different locations such as below the rotor-stator assembly (center of the bottom plate opening), top and bottom of the mixing tank (Figure 1(i)). Several locations are marked to monitor the mass fraction of tracer to investigate the mixing time and uniformity of tracer inside the mixing tank with time and rotor speed. Flow fields are also observed for different HSMs. Maximum strain rate is observed in the gap between rotor and stator and the rotor region and decreased with an increase in the radial distance. The findings provide a good estimation for further selection and development of HSMs.

References

1. Grenville, R. K., 1992. Blending of viscous Newtonian and pseudo-plastic fluids, Ph.D. dissertation, Cranfield Institute of Technology, Cranfield, Bedfordshire, England.
2. James, J., Cooke, M., Trinh, L., Hou, R., Martin, P., Kowalski, A., Rodgers, T.L. 2017. Scale-up of batch rotor-stator mixers. Part 2–mixing and emulsification. *Chem. Eng. Res. Des.* 124, 321-329.
3. Vashisth, V., Nigam, K.D.P., Kumar V., 2021. Design and development of high shear mixers: fundamentals, applications and recent progress. *Chem. Eng. Sci.* 232, 116296.
4. Vikash, Deshawar, D., Kumar V., 2017. Hydrodynamics and mixing characterization in a novel high shear mixer. *Chem. Eng. Process. Process Intensif.* 120, 57-67.
5. Zhang, C., Gu, J., Qin, H., Xu, Q., Li, W., Jia, X., Zhang, J., 2017. CFD analysis of flow pattern and power consumption for viscous fluids in in-line high shear mixers. *Chem. Eng. Res.* Des. 117, 190–204.

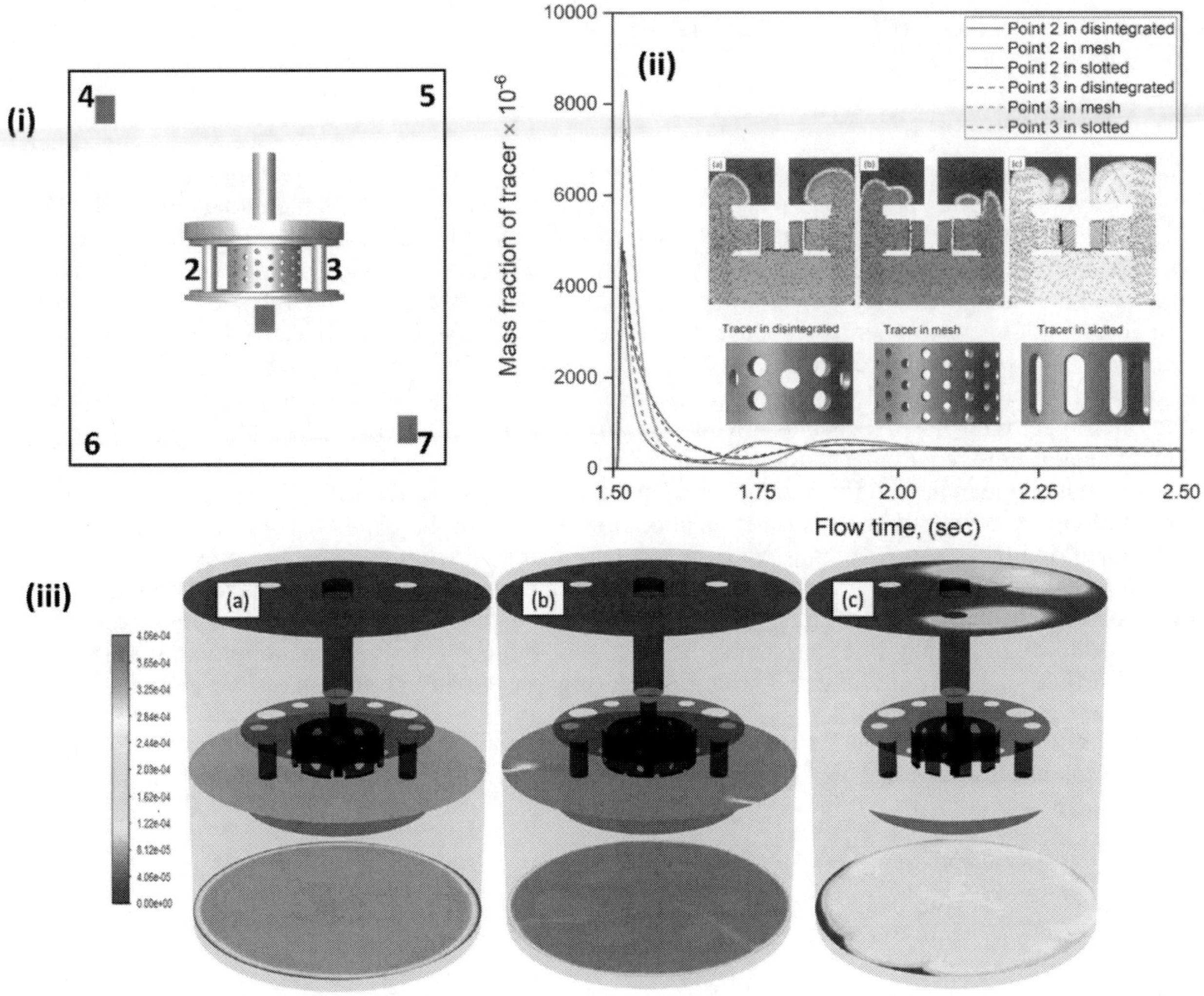
(i)
4
5
2
3
6
7
(ii)
Point 2 in disintegrated
Point 2 in mesh
Point 2 in slotted
Point 3 in disintegrated
Point 3 in mesh
Point 3 in slotted
Mass fraction of tracer × 10^-6
Flow time, (sec)
Tracer in disintegrated
Tracer in mesh
Tracer in slotted
(iii)
(a)
(b)
(c)

632635 Mixing Dynamics Characterization of Jet Mixing Reactors for Rapid Nanoparticle Synthesis

Abstract:

Purpose: Nanoparticles produced using nanoprecipitation provide a facile and rapid approach for high-throughput synthesis. Nanoprecipitation achieved in confined mixer geometries including confined impinging jets and multi-inlet vortex mixers through a process known as Flash nanoprecipitation (FNP), has been demonstrated as a scalable technology platform. Many two-inlet micromixers, such as T-mixer, Y-mixer, and confined impinging jet mixer designs, require equal inlet fluid flow rates to achieve efficient mixing and cannot operate under asymmetric flow conditions. This significantly restricts exploration of reaction parameters such as the degree of supersaturation achieved in nanoprecipitation and limits applicability for processes involving expensive reagents such as drugs. Here, we present a three-inlet jet mixing microreactor (JMR) that enables effective mixing in milliseconds and offers an opportunity to circumvent equal flow requirements of conventional micromixers. The JMR consist of two jet streams impinging on a single main line fluid stream at right angles, developing mixing in both axial and transverse directions along the exit fluid line. The JMR is about the size of a dollar coin with inlet diameters ranging 0.25-2 mm, allowing for shorter mixing times as a result of smaller length scales. The presence of a third inlet stream in the JMR enables investigation of asymmetric flow, thereby providing access to wider design space. The competitiveness of JMR over conventional stirred batch vessels and two-inlet microreactors was assessed through mixing time characterization and model block copolymer nanoparticle synthesis, with nanoparticle size and particle size distribution serving as evaluation criterion.

Research method: Mixing in the JMR was characterized using a competitive chemical reaction set, known as the Villermaux-Dushman reaction. This set consists of two parallel reactions that compete for a common limiting reagent.

A + B ---> P_1 (*Fast reaction*)

C + B ---> P_2 (*Relatively slower reaction*)

Depending on the amount of product formed from slower reaction (P_2), fluid mixing time can be estimated using established correlations. Flow dynamics were also evaluated using computational fluid dynamics, with simulations on COMSOL. Mixing dependence on different factors was explored, including dependence on fluid stream velocities, inlet diameters, main to jet fluid stream mixing ratio, and fluid viscosity. Poly(butyl acrylate)-Poly(acrylic acid) (PBA–PAA; 7500–7500 Da) served as the model block copolymer system. PBA-PAA dissolved in methanol was precipitated using water as antisolvent. Nanoparticles were purified using centrifugal filters and assessed for size using dynamic light scattering and transmission electron microscopy. Particle sizes and distributions were compared to previously published data on other micromixer systems.

Findings and Implications: Mixing time data suggest that the JMR provides rapid mixing, with timescales in range of 0.5 – 100 milliseconds that are dependent on geometry, fluid velocity, and viscosity. Compilation of mixing time data provided further insights into scaling models for the JMR. Different fluid mixing ratios were examined to explore the effect of supersaturation on polymer nanoparticle sizing, an experiment not achievable with current two-inlet micromixer systems because of equal fluid momentum requirements of opposing streams. Model polymer nanoparticle studies indicate a smaller particle size is achieved using the JMR compared to batch synthesis. Rapid mixing resulted in focused nanoparticle size (lower polydispersity), with significantly higher reproducibility compared to stirred batch vessels. Thus, JMR designs hold promise to achieve enhanced size and polydispersity control in rapid nanoprecipitation processes over batch stirred vessels and allows greater flexibility over two-inlet microreactor systems.

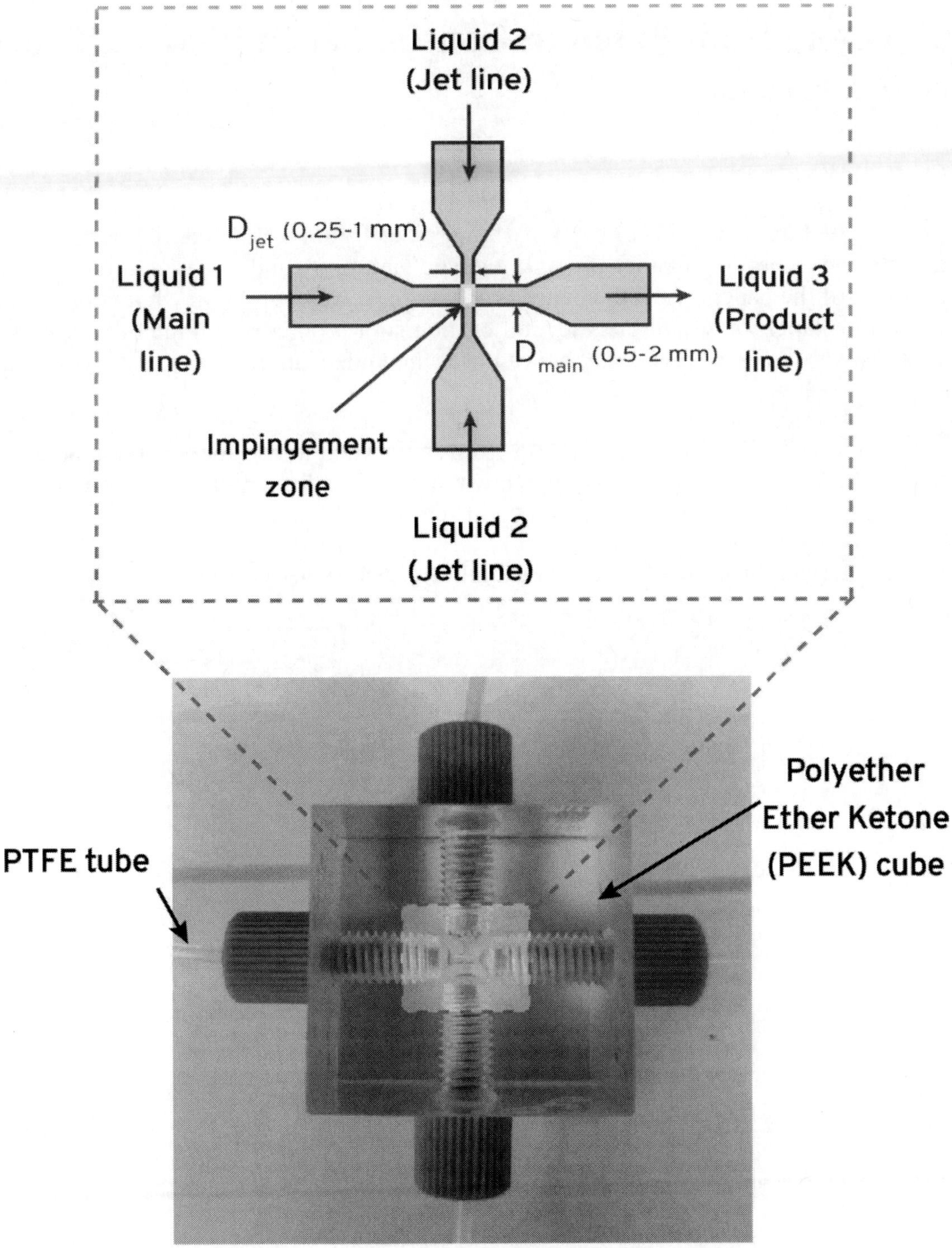

Figure 1. Jet mixing reactor

635319 Going from Batch to Continuous in Pharma Processing: Traps and Pitfalls

Abstract:

There is a great deal of interest in conversion of processes from batch to continuous as a means of lowering costs and improving quality through reduced variability and of developing new continuous processes. Most of the papers show the benefits and suggest ways of doing what is often called flow chemistry. In this paper we will discuss several of the major problems to implementing such conversions. This is not new technology but requires the understanding of the system and the devices available.

Such topics as the difficulty with transitioning from semi-batch to continuous, the amount of losses due to startup, the non-ideality of many plug flow reactors, handling highly exothermic reactions and dealing with incomplete conversion and the need for pilot plants.

626274 Application of CFD Integrated Dissolution Model in Biological Product Formulation Process

Abstract:

The dissolution process is widely used in biological product formulation development. Since a large number of compounds/excipients may be involved, a model, which is capable of predicting particle dissolution performance in selection of right compound/excipient and mixing system, is particularly desirable to minimize the amount of experimental work in biological product formulation development, scale up / down and tech transfer. A CFD integrated dissolution model was developed. The model combines CFD simulation and Noyes–Whitney equation. The energy dissipation rate of the dissolution system is linked with particle mass transfer rate through the Sherwood number, which is obtained by CFD simulation. The model was validated by the literature data. The model also can be used to evaluate effect of particle size, density, solubility and diffusivity, amount of compound/excipient, impeller size and shape, tank size, mixing speed and time on the dissolution performance. Further the scale down model workflow was developed to investigate the impact of material properties, process parameters and impeller size and shape on the dissolution behaviors in commercial scale mixing tank. Two case studies were presented: 1 evaluate impact of the particle properties (DTPA vendor change) on dissolution behavior in 4 commercial scale buffer formulation tanks and 2. evaluate impact of impeller configuration (Different size and shape impellers with/without vortex breaker) on dissolution behavior in a protein formulation tank using only 35 ml protein/vessel. The CFD integrated dissolution model and scale down model workflow were demonstrated successful in predicting dissolution behavior in biologic product commercial scale tank and supporting CAPA decision making. The simulation results also were implemented in manufacture sites.

628707 Experimental Determination and Computational Prediction of Vortex Shape and Vortex Formation in an Unbaffled Stirred Vessels Provided with a Retreat-Bladed Impeller

Abstract:

Glass-lined, stirred reactors are frequently utilized in the pharmaceutical industry to perform various unit operations such as crystallization, fermentation, cell culture, chemical reactions, and others where product contamination and chemical compatibility are concerned. The vessel walls and corresponding impeller are lined with a glass coating to prevent corrosion and provide a smooth surface for superior cleanability. Typically, this reactor is fitted with a Retreat-Blade Impeller (RBI) with low impeller clearance and a single baffle. However, in some applications, the presence of baffles may create undesirable outcomes such as particle attrition in the crystallization process, cell damage in bioreactor applications, or promote product cross-contamination. Therefore, unbaffled stirred tank are also commonly used. Although unbaffled stirred reactors may offer some advantage in some cases, such as lower power consumptions per unit volume, the lack of baffles tends to generate high swirling motion leading to a vortex formation in the impeller shaft region, primarily when operating in the turbulent flow regime. The fluid flow in a stirred vessel can be complex. Vortex shape and depth are typically dependent on the agitation speed, the liquid medium viscosity and vessel geometry. Thus, the objective of this work was to experimentally characterize vortex formation in a scaled-down unbaffled RBI system at different agitation speeds in turbulent regimes. The depth and shape of the vortex we precisely determined using a leveled laser pointer system and all experiments were conducted in triplicates to verify the reproducibility and robustness of the experimental approach. Computational Fluid Dynamics (CFD) was additionally used to predict the vortex shape and depth using a Multiple Reference Frame (MRF) model coupled with a Volume of Fluid (VOF) model. The CFD modeling was validated by comparing the predictions from the simulations with the experimental data. In all case they were found to be in close agreement. As anticipated, increasing the agitation speed resulted in a significant variation in vortex shape and depth. This result of this work can help industrial practitioners optimize pharmaceutical processes and their reactor/bioreactor design.

Experimental Result (at 200 rpm)

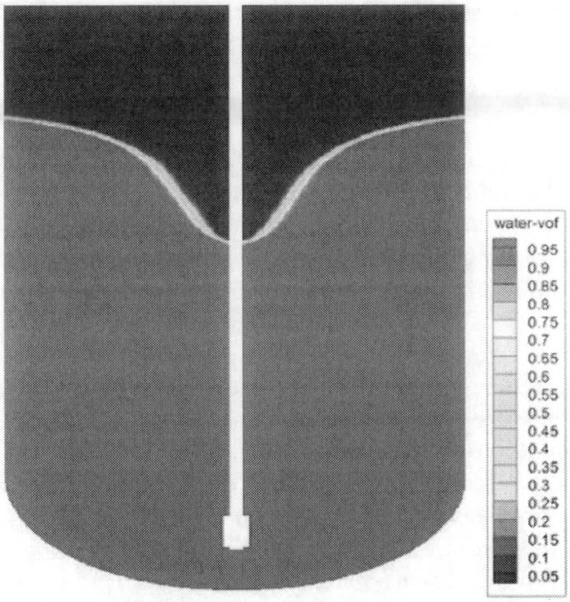

CFD Result (at 200 rpm)

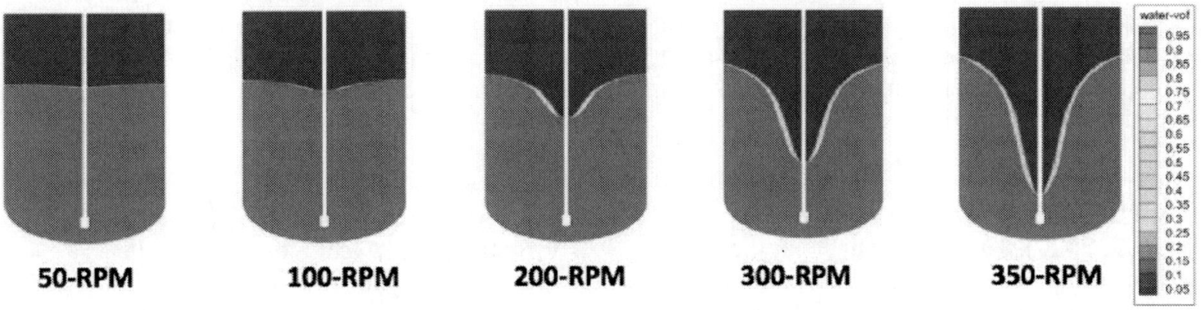

630313 Scaling Down to Scale up: Advances in Using the Bourne Protocol and Optimization in the Micromixing-Mesomixing Space

Abstract:

Reaction and crystallization processes often need to balance competing rates to achieve the desired results. Mixing can affect certain rate processes, so understanding the role of mixing is a critical part of process development and scaling. There are many potential undesirable process responses that can be affected by mixing: low yield or conversion, poor selectivity, unacceptable purity, unmanageable particle size, poor crystal morphology, inappropriate polymorph, etc. Understanding how a process responds to different levels of micromixing and mesomixing can enable more successful scaling, suggest potential process improvements, and speed troubleshooting of industrial processes. Process developers in the chemical and pharmaceutical industries need efficient ways to understand how mixing matters to their processes.

Since it was introduced in 2003, the Bourne Protocol has been used to efficiently understand the effect of impeller speed, feed time, and feed location on semi-batch or gradual-addition processes. Bourne's simple experimental plan creates a framework to understand if micromixing (mixing at the molecular or particle length scale) or mesomixing (mixing at feed plume length scale) or macromixing (mixing at the length scale of the vessel) matters in a process. Applications and enhancements to the Bourne Protocol have been discussed in a variety of venues over the past decade (see References).

Most recently, we obtained more appropriately visualized regions of acceptable process performance by plotting Bourne Protocol results in a micromixing-mesomixing space (Sarafinas, 2019 and Sarafinas, 2020). Those studies looked at examples from the literature and showed how regions of different process failures could be characterized in the micromixing-mesomixing space, enabling improved and efficient understanding of the acceptable process operating window for scale-up or scale-down. In addition, we showed approaches to make practical choices of small-scale operating conditions to mimic the micromixing and mesomixing space for large scale vessels.

The Bourne Protocol was a popular topic in the AIChE 2020 Virtual Annual Meeting, as four papers discussed different applications and advances (Sarafinas, 2020; Ataíde et al, 2020; Baudouin et al, 2020; Jordan et al, 2020). Two of the papers at the AIChE 2020 Virtual Annual Meeting discussed the importance of appropriately characterizing the local energy dissipation and local velocity in the vicinity of the feed for the proper application of the micromixing-mesomixing operating space (Sarafinas, 2020 and Jordan et al, 2020).

This paper continues the exploration of the micromixing-mesomixing space to characterize our processes. After giving a quick review on how to apply the Bourne Protocol to efficiently show how mixing matters in a process, we explore how differences between estimates of local energy dissipation and local velocity can affect proper characterization of the micromixing-mesomixing operating space. We will look at how good is "good enough" for a screening study, and how our information needs change when working on a troubleshooting or optimization study. Further literature examples and some new examples will be shown. We will consider how process practitioners can expand the Protocol to better understand their process for scaling. This approach results in achieving more robust processes and more optimal equipment designs more quickly, which should enable faster commercial success.

References:

Ataíde, F., J. Sardinha, R. Mendonça, E. Gonçalves, A. Monteiro, and S. Nascimento, "Coupling Reaction Kinetics, Calorimetry and Mixing Studies for Successful Scale-up of Exothermic Reactions," **AIChE 2020 Annual Meeting**, Virtual, November 2020.

Baudouin, O., E. Clavel, S. Dechelotte, and B. Wincure, "Modeling for the Scale-up of Processes:

Illustration on the Aspects of Heat Transfer for a Batch Reactor," **AIChE 2020 Annual Meeting**, Virtual, November 2020.

Bourne J.R., "Mixing and the Selectivity of Chemical Reactions", **Org. Process Res. Des. Dev.**; 2003; 7(4) pp. 471-508.

Jordan, C., C. Marton, Y. Agustinus, C. Wilbert, J. Perona, K. Ragazzo, and B. Holloway, "Scaling Down a Reactive Crystallization using Various Mixing Models," **AIChE 2020 Annual Meeting**, Virtual, November 2020.

Sarafinas, A., "Efficient Process Development in the Micromixing-Mesomixing Space: the Bourne Protocol and Beyond," **AIChE 2020 Annual Meeting**, Virtual, November 2020.

Sarafinas, A. "Efficiently Characterize Process Scalability in the Micromixing-Mesomixing Space Using the Bourne Protocol," **AIChE 2019 Annual Meeting**, Orlando, FL, November 2019.

Sarafinas, A. "Test Process Mixing Sensitivities Using the Bourne Protocol," **Scientific Update Webinar**, 13 November 2018.

Sarafinas, A. "Using the Bourne Protocol to test mixing sensitivities in the lab," **DynoChem Guest Webinar**, 8 August 2018.

Sarafinas, A. "A further look at the Bourne Protocol for efficient investigation of mixing sensitivities during process development," **AIChE 2017 Process Development Symposium**, Toronto, Canada, June 2017.

Sarafinas, A. and C.I. Teich in Kresta, S.M., A.W. Etchels III, D.S. Dickey, V.A. Atiemo-Obeng, **Advances in Industrial Mixing: A Companion to the Handbook of Industrial Mixing**, John Wiley and Sons, 2016.

Teich, C.I., A. Sarafinas, P.M. Morton, **AIChE 2010 Annual Meeting**, Salt Lake City, UT, November 2010 (2 papers):

- "Can this process be saved? A search for understanding using the Bourne Protocol and advanced process development tools"
- "Taking the min to the max: a case study in small scale process development using on-line reaction calorimetry and in-situ particle characterization"

Bourne Protocol and Vessel Characterization

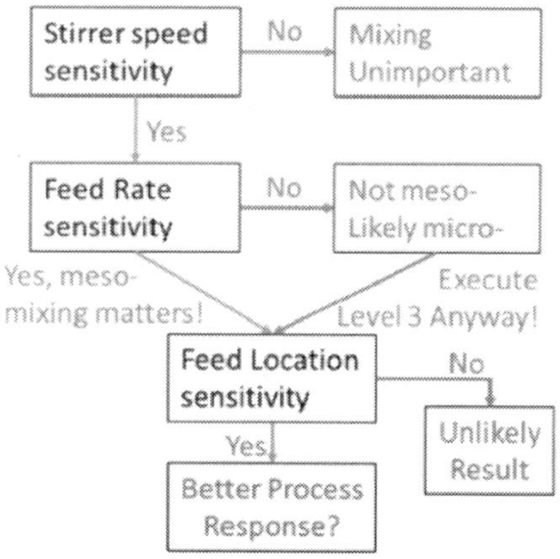

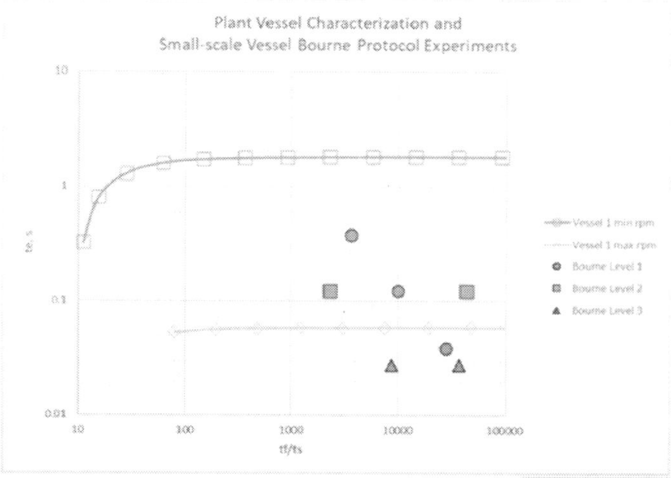

630590 Mixing Conditions for Oil-in-Water Emulsions Manufacture As Factors for Scaling-up and Product Design.

Abstract:

Cosmetics demand had shifted towards the use of organic and natural products.[1] Because of this, different tools have been proposed to simplify and accelerate the cosmetics design, including in-silico design, molecule modelling, and other approaches. In all cases, the production process and its scaling-up are factors that must be considered, especially when working with structured products (e.g., emulsions). This task can be time-consuming, resource-intensive and complex.

Several variables must be considered to ensure that product properties will be the desired ones. In the case of emulsions, equipment geometry, stirring rates, process temperatures, mixing times and heating and cooling rates, play a major impact in the final characteristics of the product. During scale-up, the influence of these variables is reinforced due to the reduction in geometrical relations such as available mixing power and volume-to-transfer-area. These variables must be adjusted during scale-up, and that is usually made by trial-and-error strategies making this process highly inefficient. Different scale-up strategies have been developed for emulsions production using dimensionless numbers in the ND^{α}.[2,3] This approach guarantees that the drop size is conserved in all scales, although we have confirmed that it is not enough to maintain all product characteristics.

In this sense, this work studied the influence of emulsification temperature, addition time, stirring rates, and cooling rate in the final characteristics of a real cosmetic emulsion (i.e., skin moisturizer) during scale-up. Experiments were accomplished at three different processing scales (i.e., 100mL, 1L and 10L) and based on the obtained results, it was found that a dimensionless number in the form of ND^{α} was conserved with α=0.5. As part of the product design approach, we analyzed the influence of each variable in the final quality of the product, and proposed some factors that must be considered during the product design process.

References

(1) L'Oréal. Cosmetics market - L'Oréal Finance: Annual Report 2018 https://www.loreal-finance.com/en/annual-report-2018/cosmetics-market-2-1/ (accessed Aug 9, 2019).

(2) May-Masnou, A.; Ribó-Besolí, J.; Porras, M.; Maestro, A.; González, C.; Gutiérrez, J. M. Scale-up Model Obtained from the Rheological Analysis of Highly Concentrated Emulsions Prepared at Three Scales. *Chem. Eng. Sci.* **2014**, *111*, 410–420. https://doi.org/10.1016/j.ces.2014.03.004.

(3) May-masnou, A.; Maestro, A.; González, C. Scale Invariants in the Preparation of Reverse High Internal Phase Ratio Emulsions. **2013**, *101*, 721–730.

631053 Real Time and Inline Monitoring of Intracellular Lipid Accumulation Using Industrial Process Microscopes

Abstract:

Microbial lipids become increasingly favorable for industrial production due to their potential to replace petroleum as a main source of fuels and chemicals as well as to provide alternative feedstock for valuable polyunsaturated fatty acids.

The present work describes the development and methodology to apply a microscopy probe for the measurement of the sizes, shapes, colors and concentrations of microorganisms and its intracellular accumulation of lipids. The in-situ measurement of these parameters is a challenge, since the monitoring instrumentation has to be applicable under various process conditions and in different process phases. High particle concentrations, different particles in multiphase systems or changing particle features and velocities are challenging for an inline measurement method.

In a first study, morphological changes and lipid accumulation in fed-batch cultivations of a heterotrophic algae Crypthecodinium cohnii were monitored. Intracellular lipid compartments are identified and correlated to the cell size and finally to the accumulation of the polyunsaturated fatty acid docosahexaenoic acid (DHA). In a second study yeast fermentations of Yarrowia lipolytica with intracellular lipid production are investigated by the same inline microscopy.

Used method

The inline microscopy is bringing the focal plane of the microscopic objective into the sample, with an optical resolution of 0,055 μm/pix. This allows object detection and semantic segmentation of yeast and its intracellular accumulated lipids. The acquired images are analyzed with an artificial neural network. The employed MASK-RCNN is a deep convolutional network. This network learns to segment images in an end to end setting [1]. Automated image analysis classifies different cells types and identifies intracellular accumulated lipids under various growth conditions.

Results

In this study, changes in the cell size distribution of the heterotrophic microalgae C. cohnii were tracked with an offline holographic microscopy and the inline photo-optical microscope. On the basis of the cell size and broadness of the size distribution, the applied inline photo-optical measurements enabled to distinguish between cells in the growth phase with little or no intracellular lipid droplet accumulation and production phase with lipid droplet accumulation. Under conditions of low growth and high fatty acid accumulation, the cell sizes and its distributions were changing accordingly. The correlation between the predicted inline measurements based on the Sauter mean diameter and the measured intracellular DHA content with GC-FID was confirmed. The results obtained by digital holographic and the side scatter values from flow cytometry, which was performed as golden standard method for single cell measurement, were in good agreement with the inline microscopy, which enables this method to photo-optically measure DHA content in real-time.

Yeast is one of the most important species in biotechnology. In this study Baker's yeast S. cerevisiae and its morphological changes during fermentation is monitored inline. The morphology of yeast cells is altering during maturation, depending on the growth rate and cultivation conditions. Inline microscopy was used to monitor such morphological changes of individual cells directly in the cell suspension. With automated image analysis it was possible to analyze budding and non-budding cells in parallel based on a trained artificial neural network. Deviations between automated and manual counting were considerably low.

The homogeneity among the population during the growth phase increased and at growth retardation, the portion of smaller cells increased due to a reduced bud formation. The maturation state of the yeast cells was determined by the budding index, which Is defined as the ratio between the number of budding cells and the total number of cells. Inline monitoring showed a linear correlation between the budding index (BI) and the growth rate during the batch cultivation. A real-time differentiation of growth activity across all process stages of several batch cultivations in complex media became feasible [2].

References

[1] Kaiming He, Georgia Gkioxari, Piotr Dollár, Ross B. Girshick, 2017, Computer Science, IEEE International Conference on Computer Vision (ICCV)

[2] Marbà-Ardébol, A.-M., Emmerich, J., Muthig, M., Neubauer, P. and Junne, S., Microbial Cell Factories, 2018, 17(1): 73.

623130 CFD Investigations of Carbopol Mixing in the Transition Region

Abstract:

Mixing is a crucial step in the production of many products, ranging from hair care to pharmaceuticals. The industrial process is typically used to blend miscible liquids, dissolve soluble compounds in a liquid, to facilitate a chemical reaction, to suspend insoluble material in liquid (e.g. paint), or to emulsify immiscible liquids (e.g. mayonnaise). Optimizing the efficiency of a mixing tank is key in product production, since not only does optimization lead to a more cost effective production process, but also a higher quality product, which is crucial in the pharmaceutical and chemical industries. To optimize the mixing process, what occurs inside the tank must be well understood. This can be unrealistic using just the physical system, as tanks can have significant volumes making modeling difficult. Computational Fluid Dynamics (CFD) offers a valuable look inside the tanks. CFD is used in modeling various physical systems to analyze velocity profiles, streamlines, volume fraction distributions, and numerous other fluid and flow properties. With data gathered from computational models, the properties and behaviors of fluid domains within a mixing tank can be studied.

The characteristics of a mixing tank are determined by the regime in which the tank operates: laminar, transition, or turbulent. Mixing is typically studied and modeled in the laminar or turbulent regime, as the equations governing these regimes are well understood. Due to the complexity of transition mixing, there is minimal modeling done in the regime despite it being a common working regime for certain industries, including the pharmaceutical, agrochemical, cosmetic, and food industries.[1-4] The lack of accurate transition mixing models hinders these industries in fully understanding their processes. The inaccuracy of transition mixing models is due to CFD's inability to predict large scale instabilities in the flow domain.[5] Not only does the inaccuracy of transition mixing models hinder industries in optimizing their processes, but it also inhibits investigation of the behavior of this regime. Processes that typically lie in the transition regime tend to involve non-Newtonian fluids due to their complex behavior. A non-Newtonian fluid is one who's viscosity is a nonlinear function of shear. This results in different Reynolds numbers throughout a tank, since the Reynolds number is a function of viscosity. The regime of a fluid domain is determined by the Reynolds number of the flow, with lower Reynolds numbers indicating laminar flow and higher Reynolds numbers indicating turbulent flow.

The goals of this project are to investigate the behavior of a non-Newtonian, shear thinning fluid, Carbopol, being mixed in the transition flow regime using CFD, and to adapt these computer-generated models to physical systems to increase the accuracy of the models. Since the transition regime spans a range of Reynolds numbers, three different ranges within the transition regime were modeled. These Reynolds numbers accurately span the transition regime of the A320 impeller, the impeller used in this study. Two sets of models were generated: (i) with changing impeller speed and (ii) with changing viscosity profile. This also allows for an investigation of the flow field's behavior with changing viscosity versus changing RPM despite having the same Reynolds number. Carbopol's viscosity can be easily modified with pH, which makes it an optimal working fluid for many industrial studies.[6] Using data from actual systems, models were generated, analyzed, and modified using commercial CFD software to accurately represent the physical systems. The models were generated using multiphase transient flow simulations with the mixture model to allow for non-

Newtonian viscosities. To compare the physical systems and the models, a portion of dyed Carbopol was added after the tanks reached steady state. Blend times and patterns were analyzed and compared across the tanks. To track the mixing in the computational model, volume fractions were used. The hope is that this work will contribute to the understanding of mixing in the transition regime.

References

(1) Rzyski, E. Mixing Time (time to homogenization) in the Transition Region of Mixing. *The Chemical Engineering Journal* **1985,** *31*, 75.

(2) Zhang, Y.; Gao, Z.; Li, Z.; Derksen, J. J. Transitional flow in a Rushton turbine stirred tank. *AIChE Journal* **2017,** *63* (8), 3610.

(3) Machado, M. B.; Bittorf, K. J.; Roussinova, V. T.; Kresta, S. M. Transition from turbulent to transitional flow in the top half of a stirred tank. *Chemical Engineering Science* **2013,** *98*, 218.

(4) Mendoza, F.; Bañales, A. L.; Cid, E.; Xuereb, C.; Poux, M.; Fletcher, D. F.; Aubin, J. Hydrodynamics in a stirred tank in the transitional flow regime. *Chemical Engineering Research and Design* **2018,** *132*, 865.

(5) Bakker, A.; Myers, K. J.; Ward, R. W.; Lee, C. K. The Laminar and Turbulent Flow Pattern of a Pitched Blade Turbine. *Trans IChemE* **1996,** *74*.

(6) Kelly, W.; Gigas, B. Using CFD to predict the behavior of power law fluids near axial-flow impellers operating in the transitional flow regime. *Chemical Engineering Science* **2003,** *58* (10), 2141.

624278 Scale-up between Tanks in Agitation of Newtonin and Non-Newtonian Fluids

Abstract:

1. ABSTRACT

Scale-up between tanks is an essential technique for the design of these industrial units. Majority of the equations available in the literature for scaling up in terms of power consumption were obtained in bench-top tanks. The present study aimed at graphical relations of scale-up based on three tanks, with a volume of 10, 50 and 750 liters, in the agitation of Newtonian and non-Newtonian liquids with an axial impeller with 4 blades inclined at 45 ° and a Rushton turbine. The results indicated a good adjustment of the experimental data with the models proposed for scale-up, which had an R^2 greater than 0.91.

2. INTRODUCTION

Tanks with mechanical impellers are equipment used for stirring and mixing liquids in various operations, such as reactors, extractors, heat exchangers, distillers, flotators and storage units. Determination of power consumption is essential for engine sizing, depending on the type of mechanical impeller, the geometry of the tank and the rheological properties of the agitation. Traditionally, the power consumed is determined from the power number as a function of the Reynolds and Froude numbers (Labik et. Al. 2018). Most industrial tanks have a volume close to 500 liters, which makes the use of the correlations proposed in experimental works and by numerical solution unreliable. Basically, the scale-up equations are obtained in the power-to-volume (P/V) between two tanks, depending on the number of Reynolds (Re) and the number of Froude (Fr), with i for the smallest tank and j for the largest tank (Equation 1).

The difficulty of Equation 1 is found with the use of non-Newtonian liquids, because in this case, there is a variation of the apparent viscosity as the rotation of the mechanical impeller is changed. For example, in the tank with the lowest volume, the flow becomes laminar and in the tank with the largest volume, it may be that the flow is in the transition region. In this situation, the best option is to scale-up through the same geometric similarity between the tank and the internals (Kresta et. Al., 2016). Therefore, an equality parameter between the tanks is chosen, such as the same number as Reynolds (Triveni et. Al., 2008). The present study aimed to present graphical relations for the scale-up between three tanks in the stirring of Newtonian and non-Newtonian liquids with the use of a Rushton turbine and an axial impeller with 4 blades inclined at 45 °.

3. MATERIAL AND METHOD

Figure 1 shows a schematic of the experimental unit used in the present article.

Acrylic tanks with volumes of 10 l, 50 l and 750 l were used, each one equipped with a Rushton turbine and an axial impeller with 4 blades inclined at 45 °. As Newtonian liquids, sucrose solution diluted 50% in water. The non-Newtonian liquid used was an aqueous solution of carboxymethylcellulose (CMC) with a mass concentration of 0.5%. The rheology of the CMC solution was analyzed on a Brookfield rheometer model DV-III, and the solution has a pseudoplastic behavior (adjusted by the power law). Each liquid was subjected to 10 rotations (N) in the range of 100 rpm to 1000 rpm depending on the volume of the tank (V) and the type of mechanical impeller, with each impeller having a diameter (Da) equal to 1/3 of the internal diameter of the tank (Dt). The tests were conducted in duplicate, totaling 360 experiments in the experimental unit. The electric motor of each of the tanks was placed on a swing on bearings, where a metallic arm was attached. At each rotation, the force (F) generated at three points on the arm (Bf) was measured using a dynamometer, so that the power (P) was calculated using Equation 2 (Rosa et. Al., 2020).

4. RESULTS AND DISCUSSION

With the results obtained in the experiments, the power consumed in all cases was calculated. The Reynolds numbers were calculated from the Metzner and Otto (1957) concept, which verified a linear relationship between the shear rate by the rotation of the mechanical impeller. In Figure 2, the results were plotted with the axial impeller with 4 blades inclined at 45 °. The ratio shown on the abscissa axis for x-axis was obtained from the ratio of the Reynolds number as a function of tanks with different volumes. The Dtj / Dti factor was placed in order to take into account the effect of the diameter of the tanks used in the study in the scale-up.

Similarly, in Figure 3, the results with the Rushton turbine are shown. It is noteworthy that the exponents of Equation 1 (disregarding the number of Froude because the tanks have Baffless) were obtained by a non-linear regression performed by combining 900 points (power and Reynolds number) by mechanical impeller. It was observed in Figure 2 and in Figure 3 an excellent adjustment of the experimental data to the model obtained. It is noted that in both cases, the scale-up relationships were initiated from the origin, which corroborates with works presented in the current literature.

5. CONCLUSION

The scale-up prediction models are valid for tanks with volumes between 10 l and 750 l, with the same geometric similarity, for stirring Newtonian liquids and non-Newtonian liquids with rheological index n between 0.79 and 1.0.

6. REFERENCE

KRESTA SM, ETCHELLS III AW, DICKEY DS, ATIEMO-UBENG VA, *Advances in industrial mixing: A Companion to the Handbook of industrial mixing*. New Jersey: Wiley, 2016.

LABIK L, PETRICRÍCEK R, MOUCHA T, BRUCATO A, CAPUTO G, GRISAFI F, SCARGIALI F, Scale-up and viscosity effects on gas-liquid mass transfer rates in unbaffled stirred tanks. *Chemical Engineering Research and Design*, v. 132, p.584-592, 2018.

METZNER AB, OTTO RE, Agitation of non-Newtonian Fluids. *AIChE J*, v. 3, p. 3-10, 1957.

ROSA VS, TORNEIROS DLM, MARANHÃO HWA, MORAES MS, TAQUEDA MES, PAIVA JL, SANTOS AR, MORAES JÚNIOR D, Heat transfer and power consumption of Newtonian and non-Newtonian liquids in stirred tanks with vertical tube baffles. *Applied Thermal Engineering*, v.176, article 115355, 2020.

TRIVENI B, VISHWANADHAM B, VENKATESHWAR S, Studies on heat transfer to Newtonian and non-Newtonian fluids in agitated vessel. *Heat and Mass Transfer*, v.44, p.1281-1288, 2008.

$$\frac{(P_j/V_j)}{(P_i/V_i)} = \left(\frac{Re_j}{Re_i}\right)^{\alpha}\left(\frac{Fr_j}{Fr_i}\right)^{\beta} \quad (1)$$

$$P = 2\pi B_f N F \quad (2)$$

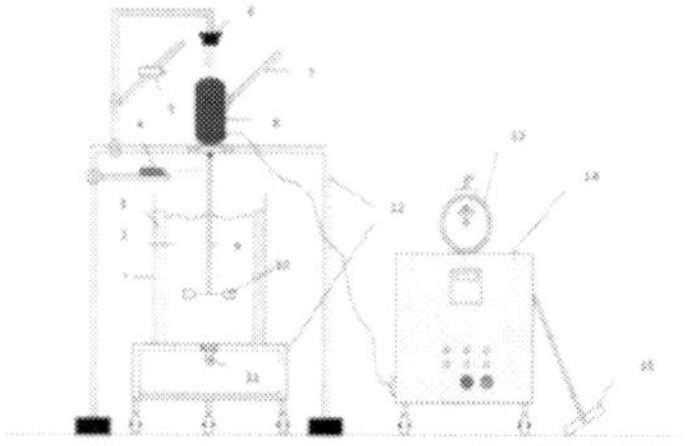

Figure 1 - Scheme of the experimental unit, being: (1) - Acrylic tank; (2) - Process fluid; (3) - Baffles; (4) - Tachometer; (5) - Digital dynamometer; (6) - Electric motor cooler; (7) - Electric motor arm; (8) - Electric motor; (9) - Impeller shaft; (10) - Rushton Turbine; (11) - Fluid drainage pipe; (12) - Support for the electric motor and tank; (13) - Analog dynamometer; (14) - Frequency inverter; (15) - Axial impeller with four blades inclined at 45°.

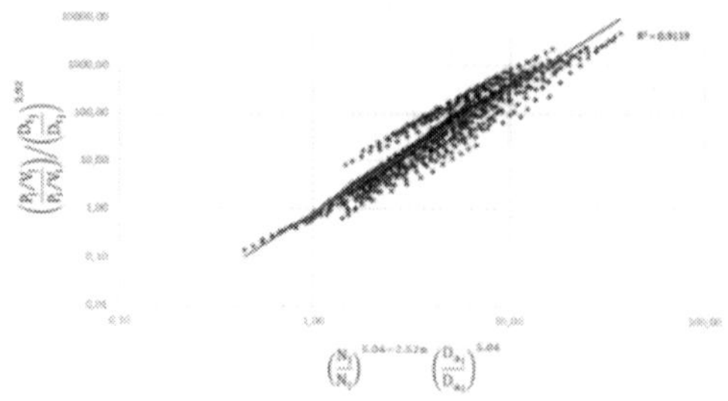

Figure 2 - Scaling ratio - Axial impeller with 4 blades inclined at 45 °

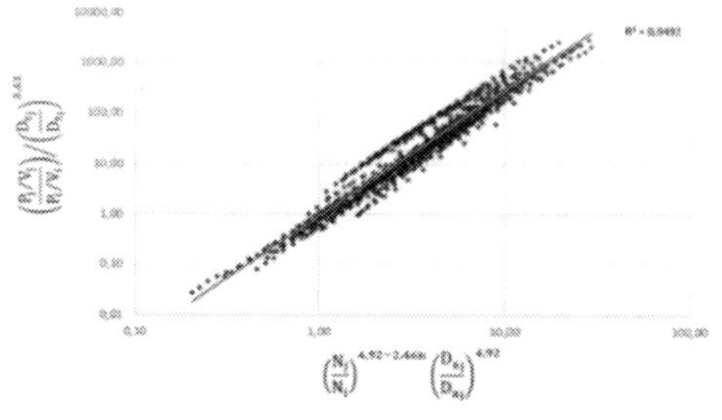

Figure 3 - Scale-up ratio - Rushton Turbine

624611 Computational Fluid Dynamics Simulation Study to Resolve Scale-up Challenges of Non-Newtonian Fluids in Pharmaceutical Industry

Abstract:

The generic pharmaceutical industry relies on the speed and rigor in the drug development process. Getting it right the first time requires detailed product and process understanding. Various unit operations like sterilization, milling, mixing, etc. govern the rapidity of the process. Mixing is one such operation, developing a proper technical understanding can yield highly productive results and can reduce experimental work thereby proving to be cost and time effective. This work provides a comprehensive approach for understanding mixing in different equipment for several parameters and scale-up strategy using computational fluid dynamics.

The first question to ask is what comprises good mixing. The judging parameters include the uniform distribution of energy with less dissipation and getting the right Critical Quality Attributes in the samples. The major challenges in mixing can be attributed to non-uniformity and unpredictability in scale-up operations. Computational fluid dynamics (CFD) is an approach that is coming into action in the pharmaceutical industry. CFD Software (ANSYS in this example) can be leveraged to understand mixing profiles and help in analyzing the technical aspects and provide a generalized methodology for the problem.

As the complexity of molecules has increased and with process intensification, the need to develop technical solutions faster and having a standardized approach will support the entire industry to reach its goal of serving people better. This study gives a comparison of equipment based on histograms, contour plots, and vector plots for shear stress, velocity, and molecular viscosity, along with the discrete phase model to draw inferences and conclusions for non-Newtonian fluid systems. Transient simulations are used to determine the particle distribution in the mixing vessel volume being dispersed from a specified location. This helps in finding time to achieve uniformity throughout the volume. Using CFD the problem at lab scale is resolved by determining the Critical Process Parameters which in turn gives us the ability to scale-up the product in a plant to achieve desired targets.

The impact of this study can be huge, as it can contribute not only to reduce experimental work at lab scale but can also improve visibility in the scale-up plant operations. It further adds as a standardized method for such issues.

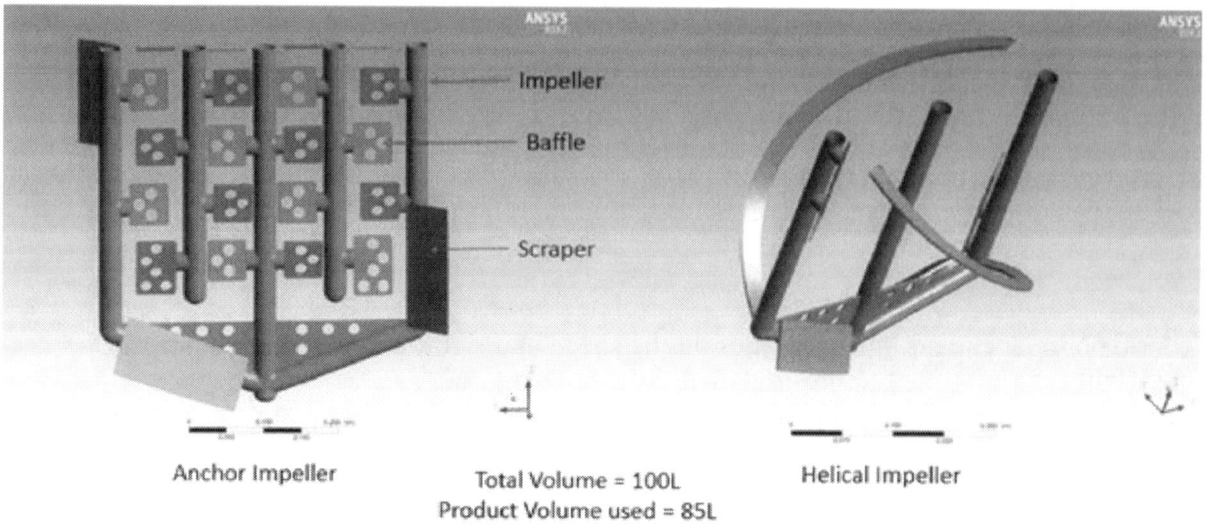

Figure 1: CAD models of Anchor and Helical impellers available at plant scale

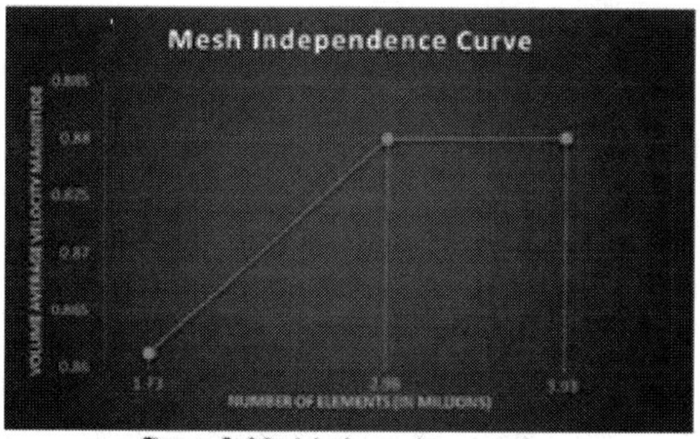

Figure 2: Mesh independence study

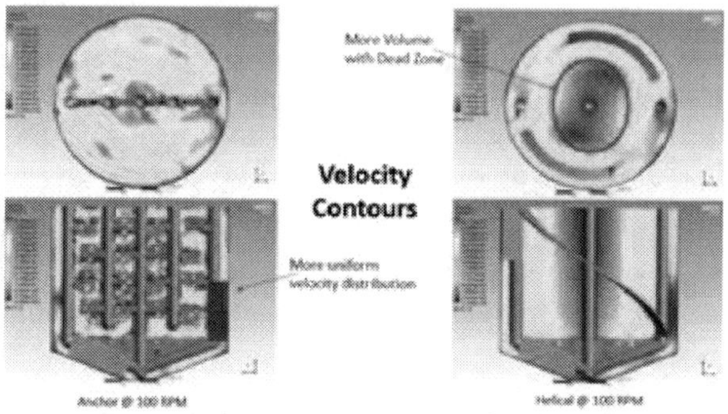

Figure 3: Velocity Contours of plant scale study

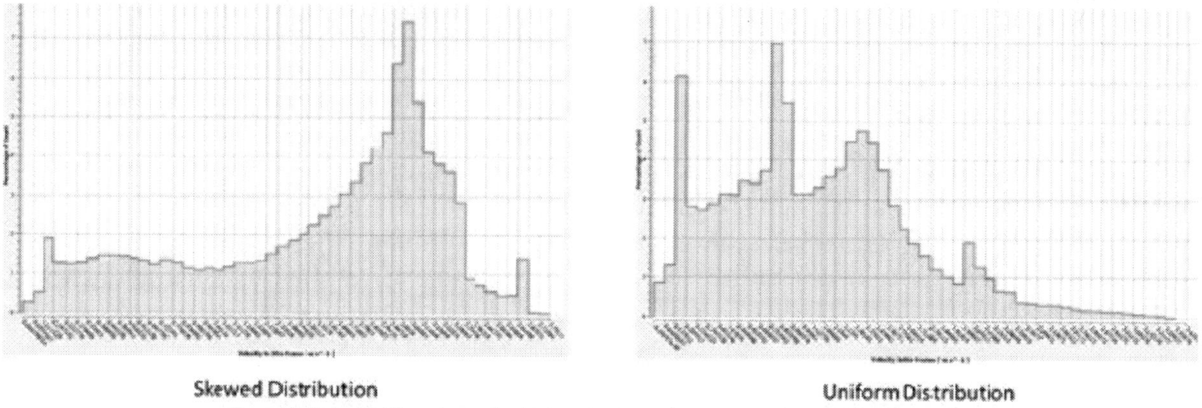

Figure 4: Velocity Histograms for Helical and Anchor impellers for plant scale study

633207 North American Mixing Forum Award Lecture: Integrating Process Engineering, Mixing and Equipment Design

Abstract:

One definition of mixing is the application of mechanical energy to produce a fluid dynamic effect which achieves a desired process result. In engineering a processing system (i.e. a vessel with mixing) other aspects of the system govern the nature of the input energy in achieving the objective. The fluid dynamics are changed by the internal vessel geometry, baffling, location of the feeds/outlets and other factors that influence the design. Other energy inputs such as the momentum in a liquid feed, gas/vapor feed or the evolution of vapor create other effects. The presence of solids and other phases can dampen the fluid dynamic effect. This paper will explore the integration of these factors, the mixing equipment design as well as other process constraints to process engineer and create a complete system for various process functions such as reactors, crystallizers, etc.

AUTHOR INDEX